Beginner's Guide to Colour Television

Gordon J. King
R.Tech.Eng., F.I.P.R.E., F.S.R.E.,
M.R.T.S., F.I.S.T.C.

Newnes Technical Books

The Butterworth Group

United Kingdom	**Butterworth & Co (Publishers) Ltd**
London	88 Kingsway, WC2B 6AB
Australia	**Butterworths Pty Ltd**
Sydney	586 Pacific Highway, Chatswood, NSW 2067
	Also at Melbourne, Brisbane, Adelaide and Perth
Canada	**Butterworth & Co (Canada) Ltd**
Toronto	2265 Midland Avenue, Scarborough, Ontario, M1P 4S1
New Zealand	**Butterworths of New Zealand**
Wellington	T & W Young Building
	77–85 Customhouse Quay, 1, CPO Box 472
South Africa	**Butterworth & Co (South Africa) (Pty) Ltd**
Durban	152–154 Gale Street
USA	**Butterworth (Publishers) Inc**
Boston	19 Cummings Park, Woburn, Mass. 01801

First published 1964 by George Newnes Ltd
Second edition 1973 by Newnes–Butterworths
Reprinted 1975
Reprinted 1977 by Newnes Technical Books
a Butterworth imprint
Reprinted 1978

ISBN 0 408 00101 1

Printed Offset Litho and bound in England
by Cox & Wyman Ltd
London, Fakenham and Reading

PREFACE

Since the original book under this title was written by Mr. T. L. Squires in 1964 dramatic developments in colour television have occurred in Europe and the United Kingdom in particular. The West German PAL system of colour television was eventually chosen which, as this book reveals, is essentially a clever refinement of the American NTSC system developed and processed by a team under the leadership of Dr. Walter Bruch of the Telefunken Laboratories in Hanover.

The PAL system differs in detail from that of the NTSC in that special encoding and reciprocal decoding artifices are employed in the chrominance channels of the transmitters and receivers in a manner to 'neutralise' displayed hue errors resulting from system phase distortion. We shall be seeing that the colouring information is carried in a subchannel and that it is the phase of the signals here which determines the absolute colours of the reproduced picture elements. The NTSC system has no such 'neutralising' feature, so that phase changes in the system produce hue changes which need to be corrected by a viewer's control—the hue control.

Moreover, the whole of the United Kingdom is rapidly becoming covered by three television programmes in colour, with provision in each reception area for a fourth programme, all the transmissions being in the u.h.f. channels and, of course, on 625 lines.

Owing to these developments I have found it necessary to change substantial amounts of original text and introduce quite a lot of new text, particularly in the chapters concerned mostly with the PAL system. Nevertheless, it has been my aim to retain the general 'down-to-earth' style of presentation and to match this to my other two Beginner's Guides to Radio and to Television. Readers requiring a more detailed insight into the servicing technicalities of colour television might find my other two books on the subject—*Colour Television Servicing* and *Newnes Colour Television Servicing Manual*—of interest.

Brixham, Devon Gordon J. King

CONTENTS

1

HISTORICAL OUTLINE

Colour television was first introduced in America in 1953, the system adopted and still employed there being that recommended by the National Television Systems Committee, called the NTSC system. This system was subsequently adopted by Japan, Canada and Mexico.

A refinement of the NTSC system was developed by the Telefunken Laboratories of Hanover, West Germany, and subsequently adopted by the German Federal Republic, the United Kingdom and by a number of other European countries. It is also being adopted in other parts of the world. The system is called PAL, which stands for *phase alternate, line*, and incorporates features in addition to those of the NTSC system for combating displayed hue errors resulting from phase distortion in the system.

A third system is called SECAM, which is an abbreviation of the French for *sequential memory system*, based on an idea by M. Henri de France. The system was launched in France in 1967, and has also been adopted by the German Democratic Republic, the USSR, Algeria, Hungary and Tunisia.

The three systems have common features, and since the PAL system is a development of the NTSC system, these two are particularly alike from basic principles.

Compatible signals

Transmitting a colour television picture would be less complicated if it were not for the problem of compatibility.

This arises because many television viewers who have no desire to purchase a colour receiver must be provided with signals which can give monochrome pictures on their ordinary black-and-white receivers. It would be both uneconomical and technically inelegant to have separate colour and monochrome systems operating side-by-side, and so television engineers have had to solve the problem of creating a signal which is capable of providing pictures in black-and-white on monochrome sets and in full colour on colour sets. This type of signal can be regarded as a 'compatible' signal in colour television parlance.

To achieve compatibility it is necessary to transmit the signal in two parts. The *luminance* part, which embodies the brightness information required by the monochrome receivers, and the 'colouring' part, called the *chrominance* signal (shortened throughout this book to *chroma*), which contains the extra information required by colour receivers. Thus, monochrome receivers employ only the luminance part of the signal, while colour receivers use the whole signal, generally termed the composite colour signal.

In each of the three systems the luminance and chroma signals are combined to form a composite signal which is then modulated onto a carrier wave for transmission. The differences between the three systems lie essentially in the method each uses in preparing the chroma signal before it is transmitted.

Chroma encoding

The chroma signal has to be accommodated in the bandwidth of an ordinary 625-line television channel provided previously for the monochrome information, and this must happen with the least interference to the monochrome reproduction. The 'interleaving' of the chroma information with the luminance information is often termed *encoding*, and this process is a function of the transmitter. The reciprocal process is *decoding*, and this is a function of the receiver.

Fortunately, since the detail in a television picture is contained essentially in the luminance information, the chroma signal needs to provide only the information required to 'brush in roughly', so to speak, the colour on the high-definition monochrome picture. The eye also assists this process significantly because its resolution to colours is well below that to white light. This has to do with the nature of the eye, where the *rods* provide the detail of the scene in monochrome and the *cones* provide the information required by the brain on the colour in the scene.

In other words, the luminance signal must provide all the information on detail over the normal monochrome bandwidth, while the chroma signal is concerned with the lower definition colouring information which engineers have been able to 'interleave' with the luminance signal with the least interaction or interference. It should be understood that the luminance signal is the equivalent of the video signal applicable to the monochrome television system. Thus monochrome receivers operating on a colour signal respond only to the luminance signal. They effectively 'reject' the chroma components of the signal.

To produce this composite and 'streamlined' signal the original colour information, which consists of three separate signals from the colour television camera, is 'encoded' into two signals which together form the chroma signal proper. The chroma signal in a compatible system must be transmitted in addition to the luminance signal to develop the main composite and compatible signal for both monochrome and colour receivers.

This book is concerned essentially with the PAL system, but since this is a development of the NTSC system, the latter will also obviously be examined, at least from first principles, and a later chapter compares this with the SECAM system.

2

COLOURS AND SIGNALS

Every colour has three important properties which are *luminance* (brightness), *hue* (colour) and *saturation* (amount of colour). In black-and-white television a single camera tube produces an electrical signal for transmission from the scene focused upon its active surface. This signal provides information on the brightness of the scene and produces in the television receiver a picture which consists of various degrees of brightness—from peak white to black. This is the *video* signal which, as mentioned in Chapter 1, is the equivalent of the luminance signal of colour television.

Colour television demands information on hue and saturation as well as on luminance, and the solitary tube of the black-and-white television camera cannot provide this extra information. The problem is solved, however, by the colour television camera which contains at least three camera tubes, one for each of the three primary colours (see later) of the scene (and nowadays sometimes one for the luminance signal alone). The scene, which is of course in colour, is first broken down into images of the three primary colours. This is done by special filters placed in front of the camera tubes. Each image is then an incomplete picture, but when the three are overlaid, as in printing coloured pictures, for example, a complete picture is formed in full colour. At this stage, however, it should be understood that each camera tube is 'seeing' only the parts of the scene appropriate to the filtered primary colour. Each camera tube thus delivers signals appropriate to the particular primary colour that it is analysing, which means that the camera system as a whole yields three primary colour signals at least. These are combined in a special way to produce the compatible signal referred to in Chapter 1. It will be realised, of course, that the signal consists of a luminance part (that required by monochrome receivers for black-and-white pictures) and a chroma part giving the information on hue and saturation as required by colour receivers for 'painting in' the colours on the black-and-white luminance and high-definition background.

Colour mixing

We know that when two colours are mixed we get a different colour. For example, we get green by mixing yellow and blue. This is called *subtractive mixing*, resulting from mixing pigments of different colours as practised by artists. The final mix is darker than the two (or more) original colours, meaning that some brightness has been lost. It is for this reason that when an artist wants a high key he uses a 'sparkling' white surface on which to begin his work—that is, he has to start with as much brightness as possible because he is losing it as he lays on and mixes colours.

Colour television uses a different kind of mixing to provide the various hues, called *additive mixing*. This differs from the subtractive mixing of pigments since lights instead of pigments are combined. With additive mixing brightness is not lost, and by using three colours with lower individual brightnesses it is possible to obtain white light—a situation which is not possible by the use of pigments (e.g., paints).

Figure 2.1 shows the result of mixing the lights from two torches after passing through red and green filters. It will be seen that where the red light overlaps the green light on the screen a new colour light is produced which, in this particular case, is yellow.

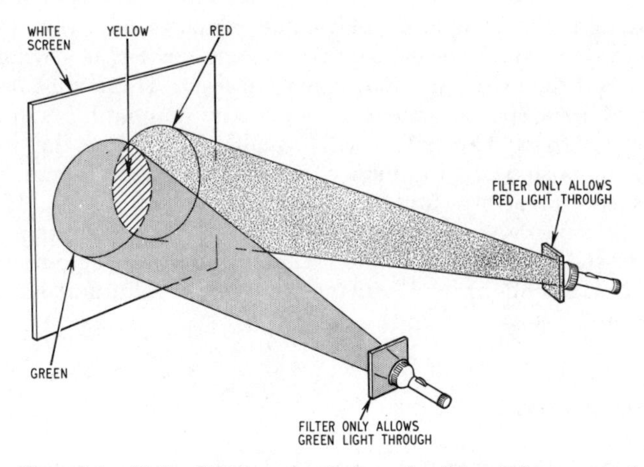

Figure 2.1. Yellow light is produced when red and green lights are caused to overlap on a white screen

All light has two important characteristics; these are colour and intensity. In physical terms, colour or hue is related to the wavelength and hence frequency in the colour spectrum. The light spectrum is a small part of the electromagnetic wave spectrum, which includes the waves we use for radio and television transmissions and those of x-rays, gamma-rays, cosmic rays and the like. The only difference between light waves and those from other parts of the overall electromagnetic wave spectrum is that of wavelength and hence frequency.

The visible light spectrum ranges from 780 nm at the red end to 380 nm at the violet end, while radio waves start at about 3 000 m and cosmic rays go on to about 3×10^{-17} m at the other end of the spectrum. One nanometre (1 nm) incidentally, is equal to 10^{-9} m.

Figure 2.2 gives an impression of the energy distribution curves for red, yellow and green lights. It is the mixture of the broken-line curves for red and green which gives the eye

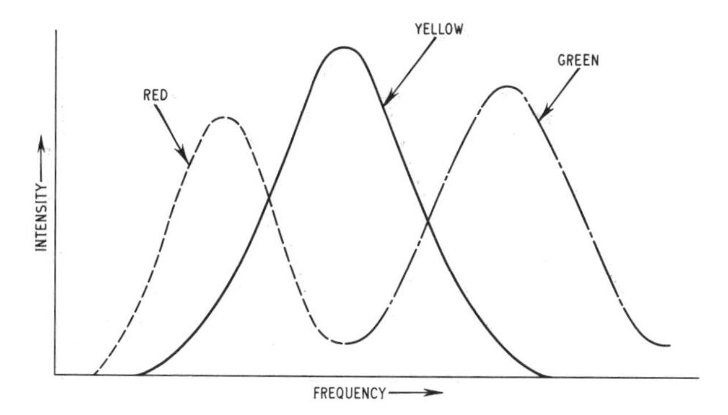

Figure 2.2. Elementary energy distribution curves for red, yellow and green lights

the same colour impression as the full-line curve for yellow. In other words, the eye integrates the red and green lights to discern yellow light.

Television primaries

The additive primary colours used in television are red, green and blue. If lights of these colours are mixed in the correct proportions the eye will see white light. The idea is illustrated by the three torches in *Figure 2.3*. This obviously implies that white light is composed of lights of the three primary colours—red, green and blue.

One way of looking at this is in terms of the light from the sun, which is very much white light. However, when this light is refracted by passing through a cloud of moisture or through a prism the component colours are displaced one from the other, and we obtain the typical rainbow effect, in which the predominant colours are red, green and blue (with, of course, intermediate hues). By using some optical device to combine the rainbow of colours we should end up with the original white light!

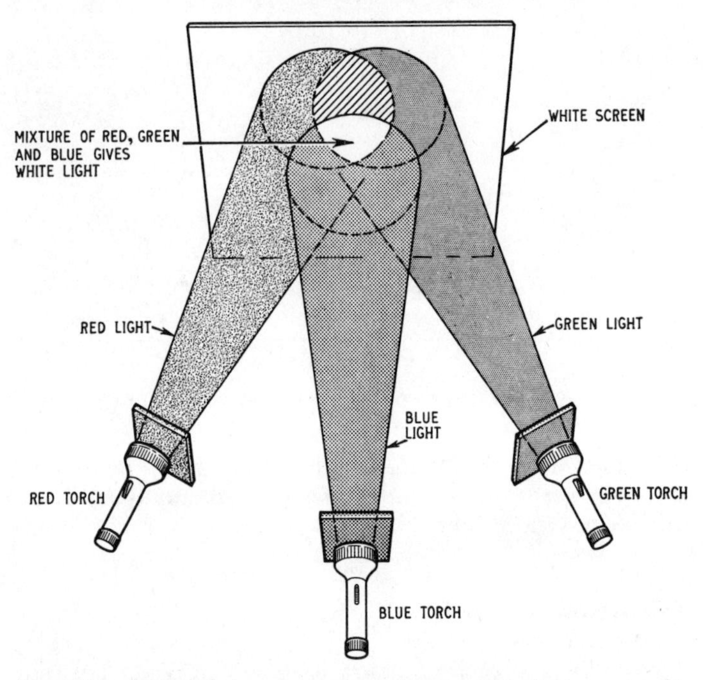

Figure 2.3. White light is produced when red, green and blue lights are caused to overlap on a white screen. The nature of the white light depends on the intensities of the red, green and blue lights. Equal-energy white light is produced when the red, green and blue lights are of equal energy. This is a hypothetical 'white' used in colour studies. White light for colour television is called Illuminant D, which simulates direct sunlight and north-sky light mixture

It will be apparent, therefore, that by using various proportions of the three primary colours in terms of lights it is possible to produce virtually any colour.

Complementary colours

Since we can regard white light as consisting of red, green and blue lights, it is now fairly obvious that by removing the blue light (leaving only the red and green lights) we would end up with yellow light, as in *Figure 2.1*. Yellow is a complementary hue. It is, in fact, complementary to blue since blue was the additive primary which had to be removed from white light to produce it. By similar tokens the complementaries of red and green are cyan and magenta, which means that cyan is produced by the addition of green and blue and magenta by the addition of red and blue.

Thus we have the primaries red, green and blue and the complementaries cyan, magenta and yellow; all these colours are obtainable—with white light—from the three primary colour lights.

It is difficult to visualise the wide range of hues that can be obtained from the three television primaries by changing their relative intensities; but those who have been viewing a good-quality colour television picture under proper ambient lighting conditions for any length of time will appreciate that almost the full range of natural colours can be simulated.

At this juncture it is noteworthy that the saturation of any colour is at maximum with the least white light added to it. Thus any displayed colour can be desaturated merely by turning on more white light (e.g., lights of red, green and blue in correct proportions). Moreover, desaturated colour will result from viewing a colour television receiver under conditions of high ambient lighting.

Returning briefly to the subtractive colour mixing, a pigment which the eye discerns as yellow when white light is falling upon it reflects both the red and the green from the white light and absorbs the blue light. Thus the eye discerns yellow because it integrates the red and green reflections.

What we have seen so far, then, is that signals are produced by the colour television camera of the primary colours and of the luminance of the scene. Should the requirement be for a colour-only television system—devoid of compatibility—then electrical signals equivalent only to the three primary colours would need to be transmitted. These could be obtained from the scene by an arrangement of filters and then transmitted in a relatively simple manner so that the receiver would reassemble them in the form of overlaid red, green and blue pictures (as in colour printing). Synchronising would be required as in monochrome television and for the chroma signals, ensuring that the red, green and blue components occur on the complete picture at the right places. The need for compatibility, though, tends to complicate the system as we shall see later.

Colour filters

When a camera tube is presented with a focused image of the scene being televised it responds only to the luminance of the image or scene. For example, the same electrical output would result from a red, green or blue component of a *given* reflective light energy. In practice, of course, the reflected light energy differs between the colours in the scene, so that they are delivered as different shades of grey from a monochrome receiver.

A colourless scene also produces a signal output related to the amount of light that it reflects. Pure white reflects most light so that the resulting bright image at the camera tube produces a high (peak white) signal level. Total black reflects zero light and hence fails to produce an image at the camera tube, so there is zero signal output. The 'greys' between peak white and total black produce signal levels between that corresponding to peak white and zero, corresponding to black.

To summarise, the luminance output due to the standard colour bars corresponds to six shades of grey between black and white, with blue being the darkest shade (next to black) and yellow being the lightest shade (next to white)—see also under *Luminance signal*, later.

(The standard colour bars have sections of black, blue, red, magenta, green, cyan, yellow and white, and in monochrome are displayed as a grey scale between black and white.)

To televise in colour it is first necessary for the camera to analyse the colour content of the scene in terms of the three additive primaries. This colour analysis is achieved not by the camera tubes alone, but by a series of colour-selective filters which pass only one primary colour, rejecting the others. Each filter guides the amount of its particular colour present in the scene to a single camera tube, which then assesses the brightness of that colour and yields an electrical output equivalent to the amount of energy present in the scene for that colour alone. The colour television camera thus employs three camera tubes for this purpose, one for each primary colour with the appropriate filter in front of it, instead of the solitary tube of the monochrome television camera.

To obtain a black-and-white picture from a colour television camera it is necessary for the three camera-tube outputs to be combined, and this is in fact what is done in the compatible system.

Elementary colour television system

An elementary colour television system is depicted in *Figure 2.4*. Here we have the three camera tubes and the red, green and blue filters fed with light from the scene via three separate paths, achieved by silvered mirrors and dichroic mirrors. It will be understood, of course, that the three paths of light are identical in colour content, the colour filtering being done by the filters appropriate to the three camera tubes. These remove all except the wanted primary colour —red in the case of the first tube, green for the centre tube and blue for the other tube shown. Each camera tube thus has an image on its active surface composed of one primary colour part of the scene.

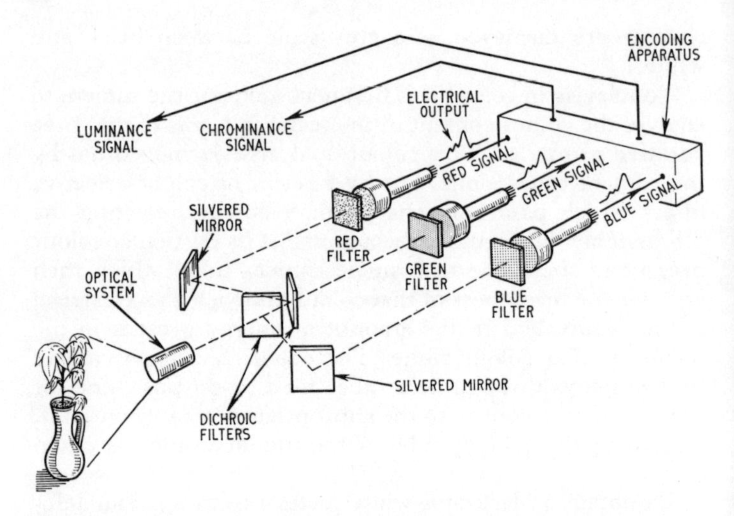

Figure 2.4. Basic elements of the front end of a colour-transmitting system

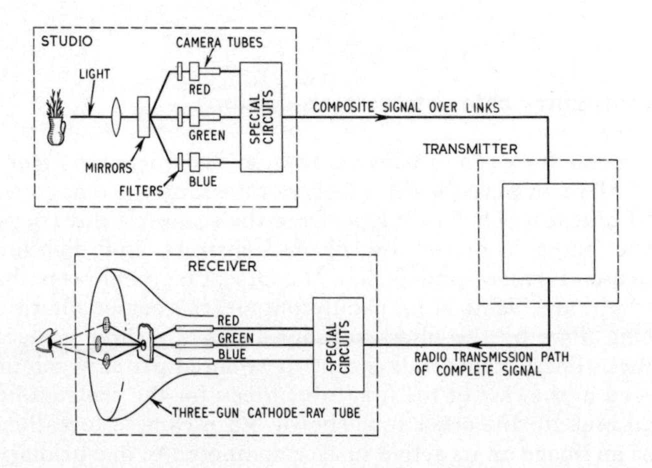

Figure 2.5. Elementary concept of a complete colour television system

Figure 2.6. Marconi Mk. VIII automatic colour television camera. Although weighing less than 49 kg (110 lb) with lens, this first-in-the-field camera features automatic line-up, automatic colour balance and automatic iris control for standby operation. Facilities include tilting viewfinder, manual or servo zoom lenses (10 : 1, 15 : 1 or 16 : 1) and remote-controlled filter turret. Sensitivity ranges from 50 lux to 800 lux at f4 (by courtesy of The Marconi Company Ltd.)

The output from each camera tube is thus the electrical equivalent of the amount of each primary colour present in the scene. These outputs are then processed to produce the luminance and chroma signals.

Luminance signal

The luminance signal (that which carries the information on brightness) is obtained by adding the red, green and blue signals from the camera tubes in the respective proportions of 30, 59 and 11 per cent, making a total of 100 per cent. This is best observed in simple algebraic terms such that the luminance signal, denoted Y in colour television, is

$$Y = 0.3R + 0.59G + 0.11B$$

Here R, G and B correspond to the primary colours of red, green and blue. If each primary colour signal from the corresponding camera tube is initially adjusted on a pure peak white scene for 1 V, then by taking 30 per cent of the signal from the red tube, 59 per cent of the signal from the green tube and 11 per cent of the signal from the blue tube and adding them all together we would obtain a total of 1 V Y signal. The tubes, of course, are not red, green and blue. The colours merely indicate the tubes which are filtered likewise. In colour television we often talk of red signals, blue electron beams and green guns and tubes, for example, not because they are physically coloured but as a colloquial identification of the circuit, component, channel, etc. which is concerned with that particular colour. Such a mode of expression is adopted throughout this book.

It might be wondered why equal proportions of red, green and blue signals are not combined to produce the luminance signal. The reason for this is that the human eye fails to respond equally to all colours. It is more sensitive to green than to red and more sensitive to red than to blue. Hence the use of the proportions previously mentioned.

It is important to remember that the luminance signal is in general character the same as that termed video signal in the monochrome system. In fact, from first principles a monochrome television camera (with a single camera tube) adjusted to yield 1 V video signal from a pure white input would produce 0.3 V video signal from a fully saturated red input, 0.59 V from a fully saturated green input and 0.11 V from a fully saturated blue input—the same proportions as given from the camera tubes of a colour camera scanning a pure white scene, the three adding to produce 'unity' Y signal.

Clearly, then, a monochrome receiver with a video input comprising the Y signal obtained from a colour television camera would produce a picture in black and white which would hardly be distinguishable from that obtained from a single-tube monochrome television camera. Here is one practical aspect of compatibility. (Gamma correction produces mild brightness errors on monochrome receivers in areas corresponding to colour.)

Colour-difference signals

We are now aware that a coloured scene possesses three important characteristics. One is the brightness of any part of it, already discussed as luminance, two is the hue of any part, which is the actual colour, and three is the saturation of the colour; that is the depth of colour. Any hue can vary between very pale and very deep, and the degree of the colour is basically its saturation. For example, a very deep red would be highly saturated, while orange is also red (that is, a similar colour wavelength) but less saturated, which means that white light has been added to it.

When a camera tube is scanning the red in a scene, as an example, it receives information on the luminance, hue and saturation because all three are obviously present in any colour. However, remember that the luminance signal is processed separately by proportioned addition of the primary colour signals and also transmitted effectively in 'isolation' so that it can be used by monochrome receivers. This, then, means that an additional signal has to be added to and transmitted with the luminance signal so that colour receivers will obtain the extra information required on the hue and saturation of the picture elements. This signal is called the chroma signal.

It is formed initially on a subcarrier which is then suppressed at the transmitter and reclaimed at the receiver, as we shall see later. The subcarrier is modulated in a special way by colour-difference signals, of which there are three but only two of them need to be transmitted.

The three colour-difference signals are red minus the luminance signal, green minus the luminance signal and blue minus the luminance signal. By red, green and blue is meant the primary colour signals delivered by the colour television camera, while the luminance signal is the Y signal as defined by the expression given earlier.

Thus, in simple algebraic terms the three colour-difference signals are:

$$R-Y, \; B-Y \text{ and } G-Y$$

It is not necessary for the subcarrier to be modulated with the three colour-difference signals, and provided two are used the third can easily be recovered at the receiver. It has been found technically convenient to recover the G−Y at the receiver and to use the R−Y and B−Y signals for modulating the subcarrier. We shall see later that in order to get the two colour-difference signals to modulate the subcarrier in a manner that permits the reclaiming of the two signals at the receiver with the least interaction, a special kind of modulation is necessary, called *quadrature modulation.*

This, then, means that at the receiver, subsequent to the colour decoding, the R−Y and B−Y signals are obtained and from these the missing G−Y signal is reclaimed.

Chroma signals

In some literature the colour-difference signals themselves are called the chroma signals. This is not strictly true, however, since the colour-difference signals, although carrying colouring information, are effectively in video-frequency form. The chroma signal is really a product of the subcarrier and the R−Y and B−Y quadrature modulation. There is another important component, called the *colour bursts.*

These are concerned essentially with 'synchronising' the receiver's decoder to the subcarrier suppressed at the transmitter so that the R—Y and B—Y signals can be demodulated in isolation.

The chroma signal proper, then starts life at the transmitter as a special subcarrier upon which are amplitude modulated the R—Y and B—Y colour-difference signals, the modulator being of a type that deletes the subcarrier itself so that only the sidebands of the R—Y and B—Y signals centred on the subcarrier frequency are integrated to form the real chroma signal. At the receiver, therefore, the subcarrier has to be produced by a generator and accurately synchronised to the subcarrier suppressed at the transmitter in order to secure correct demodulation. This is where the colour bursts come in since they are 'samples' of the real subcarrier at the transmitter. The colour bursts occur line-by-line, being located on the back porches of the line sync pulses. Their nominal phase and hence frequency correspond exactly to those of the subcarrier. Each colour burst consists of about ten cycles of sine wave signal as shown in *Figure 2.7*.

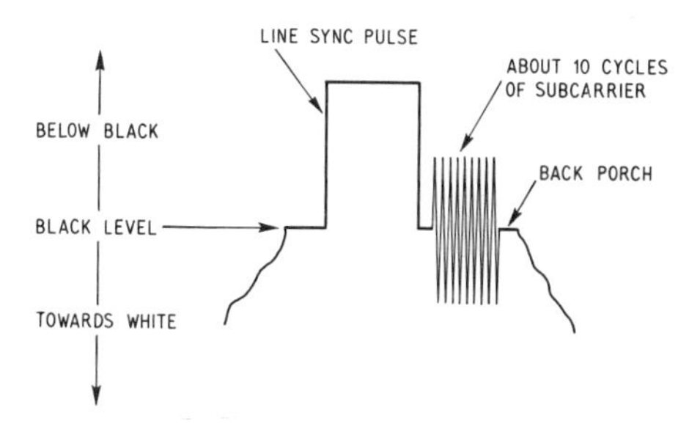

Figure 2.7. Showing a colour burst on the back porch of the line sync pulse. Frequency and phase correspond to the subcarrier and each burst consists of about ten cycles of sine wave signal

It is interesting to observe that the colour-difference signals fall to zero amplitude when the televised scene is devoid of colour—that is, when greys and whites are being transmitted. This is an important feature of contemporary colour television since it constitutes a significant aid towards compatibility and should thus be well understood.

We have seen that the red, green and blue signals from the tubes of a colour television camera can be conveniently tailored to 'unity' (1 V) on a pure peak white input. Thus $R = 1$, $G = 1$ and $B = 1$. Y, we have seen, is equal to $0.3R + 0.59G + 0.11B$, which means that on a pure peak white input we have $0.3(1) + 0.59(1) + 0.11(1)$, which equals 1. Clearly, then, from this we get $R - Y = 1 - 1 = 0$ and $B - Y = 1 - 1 = 0$. The same conditions exist on greys when the red, green and blue signals from the tubes are less than 'unity' but still equal. On colour scenes, of course, the red, green and blue signals are not equal and so colour-difference signals arise, and only when this happens is chroma signal produced.

It is possible, of course, to calculate both the luminance signal and the colour-difference signals from the colour scanned by the colour television camera, remembering that anything below full saturation means that white is added to the predominant hue in terms of the three primary colours in the proportions required for the luminance signal. Thus, while the red, green and blue signals become unequal when the camera is scanning a coloured scene, the Y signal still retains the proportions of 0.3, 0.59 and 0.11 of the red, green and blue signals.

For example, purple of below full saturation comprises a mixture of red and blue with a little green signal too, so that the Y proportions of the red, green and blue provide the 'white' which reduces the saturation. Thus we may have $R = 0.6$, $G = 0.1$ and $B = 0.5$, meaning that the luminance signal Y is equal to $0.3(0.6) + 0.59(0.1) + 0.11(0.5)$, or $0.18 + 0.59 + 0.55 = 0.294$. Using this for Y, therefore, $R - Y$ is $0.6 - 0.294 - 0.306$ and $B - Y$ is $0.5 - 0.294 = 0.206$. When the Y signal is of a greater voltage than the primary colour components of the colour-difference signal, the colour-difference signal as a whole then assumes a negative value as, of course, would be expected. The colour-difference signals, therefore, can swing from zero to a maximum both in the positive and negative directions. We shall see later that the colour-difference signals modulate the beams of the colour picture tube.

Constant luminance principle

The colour system so far described is essentially applicable both to NTSC and PAL, where the luminance information is transmitted as amplitude modulation of the vision carrier, while the two colour-difference signals are transmitted as quadrature amplitude modulation in a subchannel in which the subcarrier is suppressed.

Because the luminance signal carries only the brightness information and the chroma signal only the colouring information, the term *constant luminance* is commonly used to describe the scheme. It ensures, of course, that a monochrome receiver will give a display in black and white of a colour transmission. This is because the colouring information is disregarded by a monochrome receiver. However, a monochrome display of a colour transmission is not entirely panchromatic—e.g., areas of high saturation fail to correspond to the exact shades of grey as would be obtained from a truly monochrome transmission.

This, resulting from the effects of gamma correction (page 74) at the transmitter, necessary to equalise for the non-linear input/output characteristic of the picture tube, constitutes partial failure of the constant luminance principle. The coloured areas of the monochrome display are not quite as bright as they should be—but subcarrier rectification by the picture tube fortunately tends to combat the effect. In general, however, the partial failure of the constant luminance principle does little to detract from the quality of the picture.

PAL and SECAM differences

We shall be seeing in Chapters 4 and 5 that the PAL system differs from the NTSC system in that the R – Y component of the chroma signal is transmitted with a phase reversal on alternate lines. In this system the R – Y component of the chroma signal is called the V chroma component and the B – Y component the U chroma component, though the terms V signal or V chroma signal and U signal or U chroma signal are not uncommonly used. The V and U signals are obtained by *weighting* of the R – Y and B – Y signals respectively prior to them being used to modulate the sub-carrier.

In the NTSC system the R – Y and B – Y components of the chroma signal are modified to I and Q signals respectively. These differ from the V and U signals of the PAL system both in terms of general weighting and orientation with the respect to the chroma axes. With PAL the R—Y component lies along the V axis and the B—Y component lies along the U axis, as we have seen. However, with NTSC one of the two chroma components is shifted so that it lies along the I axes, which is that of maximum colour acuity, while the other lies along the Q axis which, of course, must be located in phase quadrature with the I axis (this will become clearer later).

Moreover, the I signal is given a greater bandwidth than the Q signal because it is associated with hues over the orange–cyan range to which the eye is most able to resolve. The purple–yellow–green hues associated with the Q signal are less well defined by the eye and can thus be handled within a smaller bandwidth. With the PAL system the chroma bandwidth is the same for both the V and U signals.

Although the SECAM system uses the two NTSC-type colour-difference signals, but based on the R – Y and B – Y axes, the scheme differs significantly from the main NTSC features. For one thing the colour-difference signals are *frequency modulated* upon the subcarrier; and for another the subcarrier is modulated by the R−Y signal on one line and by the B−Y signal on the next line, and so on. Thus the R−Y and B−Y components are not present in the chroma signal simultaneously, as they are in the NTSC and PAL systems. Moreover, the subcarrier is not suppressed at the transmitter, which means that components of the chroma signal are present in grey as well as in coloured picture areas.

More basic details are given on the NTSC and PAL systems in Chapters 4 and 5 and the SECAM system in Chapter 10.

3

PICTURES AND SIGNALS

This chapter describes the main items of equipment used in the studio to create a colour television signal.

The light rays from the scene are focused onto the active surfaces of the three camera tubes, via the red, green and blue filters, so that the images represent the red, green and blue components of the scene being televised. Each image produces in each tube an electric pattern which is a replica of the optical image on its active surface. As with ordinary monochrome television, the electrical pattern then undergoes a process called *scanning*.

Scanning

The problem is first to convert the two-dimensional pattern into a single dimension, which can be likened to reading a book. The eye travels along each line from left to right and then swiftly back again from right to left to read the next line, and so on. It is thus also moving down the page. This natural process is tantamount to television scanning of a two-dimensional pattern which converts it to a single-dimension spread-out in time.

A similar function occurs in television. The two-dimensional image on each camera tube is 'read' by an electron beam inside the tube which is arranged to scan the focused image from left to right, to go quickly back to the left again for a return scan and so on, concurrently with the electron beam being caused to move less speedily downwards from the top of the image to the bottom.

On each horizontal scan, called a *line scan*, the electron beam precipitates a series of varying voltages corresponding to the amount of light due to the image at each point on the tube surface being scanned at any instant.

A scanned image is thus composed of a given number of lines which, of course, depends on the line scan period and on the relative rate at which the electron beam is moving downwards. The scan from the top to the bottom of the image is called a *field scan*.

A 625-line television picture implies that there are 625 line scans per image or *frame* as a complete picture is called. In practice there are slightly fewer than this owing to the necessary intervals between each top-to-bottom (field) scan when the electron beam has to return very quickly from the bottom to the top to start the next field scan. During these intervals various pulses occur for synchronising the field scans at the receiver with those at the transmitter and for 'stabilising' certain aspects of the picture signal. While these pulses are occurring there is effective line scanning but no actual picture information. Hence some of the lines are blacked out.

There are also intervals between each left-to-right (line) scan when the electron beam has to return swiftly from the right to the left to start the next line scan. During these intervals the line synchronising pulses, colour bursts and black level intervals (e.g., front and back porches to the line sync pulses) occur—see *Figure 2.7*.

The speedy field and line scan returns are called '*flybacks*' or *retraces*. They occur, of course, much more quickly than the line and field scans themselves. Thus the image is converted into a whole series of electrical signals with intervals between each line and intervals between each field.

The main item of studio equipment is the colour camera. This may contain three or four camera tubes. When there are four, one is used for each primary colour and a separate one for the Y (luminance) signal. When there are three, the Y signal is derived by the addition of suitable proportions of red, green and blue signals, as explained in Chapter 2.

Camera tubes

Early cameras employed photoemissive tubes called image orthicons. The latest colour cameras, however, employ photoconductive tubes sometimes called vidicons.

Although the image orthicon tube has departed from colour television a brief description of its operation should really be included here for the sake of completeness if not for academic reasons.

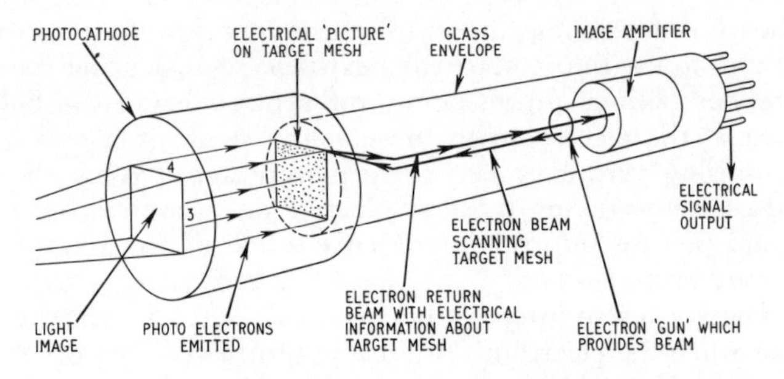

Figure 3.1. Primary elements of image orthicon camera tube. Tubes now in use are of the photoconductive type (Figure 3.2)

Referring to the simplified diagram in *Figure 3.1*, the image of the scene to be televised is focused onto the photocathode by a suitable optical system (see *Figure 2.6*). Under the influence of this image, the photocathode emits electrons. The quantity of electrons emitted depends on the brightness of the various parts of the image, the bright parts producing more electrons than the less bright parts.

The electrons are attracted to the target where they result

in electric charges corresponding to the brightness values of the image. This 'charge pattern', as it can be described, changes as the scene changes. At the other end of the tube an electron gun produces a beam of electrons which is deflected vertically and horizontally as already described so that it scans the target area and hence the charge pattern. The beam thus samples each part of the charge pattern and, depending on the magnitude of the charge, some electrons of the beam reverse, forming a return beam, while others remain on the target and neutralise the electrons produced by the photocathode. The net result of all this is that the return beam reflects the charge pattern in terms of varying strength (e.g., quantity of electrons). For example, a strong light on the photocathode would result in a relatively large charge on the corresponding part of the target, so that when it is scanned by the beam there is a relatively large change in the electron content of the return beam.

The return beam passes through an image amplifier which increases the strength of the scanned information. The tube thus provides a sequential series of electrical signals corresponding in value to the brightness of the various parts of the image.

The action is the same for all the tubes, and the three primary colour signals are directed to the encoding and synchronising equipment which prepares them with the other signal components for transmission as a compatible signal.

Photoconductive camera tube

The basic parts of the photoconductive vidicon camera tube are shown in *Figure 3.2*. The three primary parts are the electron gun, the scanning system and the target upon which an image of the televised scene is focused, as in the orthicon.

The electron gun works in a similar manner to that in a display tube (see *Beginner's Guide to Television*). The beam electrons are brought to a high velocity by a system of anodes, though the velocity here does not need to be as high as that required for the electrons in the display tube, the kinetic energy of which are exchanged for screen illumination. While a display tube (colour) may have some 25 kV on the final anode, the accelerating potential of the vidicon is generally measured in hundreds instead of thousands of volts.

The beam is focused electromagnetically in some tubes so that it impinges as a very small 'point' on the target area in the interests of resolution. Deflection is by field and line

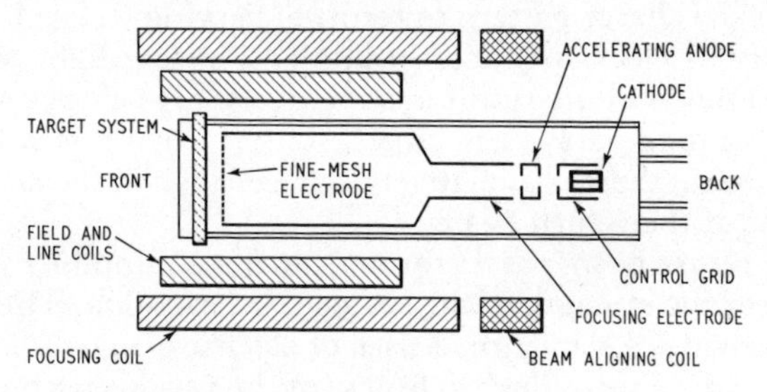

Figure 3.2. Primary element of photoconductive vidicon camera tube

scanning coils. The beam alignment coil shown in *Figure 3.2* provides an adjustment to correct minor tolerances in the tube's construction. There is also a focusing electrode in the tube illustrated, which is used in conjunction with the magnetic focusing field. Basically, the principles with regard to the scanning are virtually equivalent to those of the display tube, though of course points of detail differ.

Figure 3.3 shows the target end of the tube. The target proper consists of a photoconductive material which exhibits a very high resistance (called the dark resistance) when the intensity of the light falling upon it is very low, the resistance reducing as the light intensity increases. A lens system focuses the image of the scene upon the target area as shown in *Figure 3.3*.

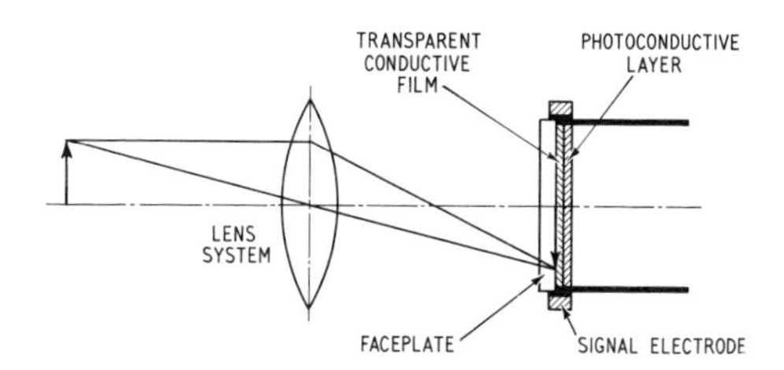

Figure 3.3. Target area and lenses arrangement of the vidicon camera tube

Due to the action of the scanning by the beam the target area can be regarded as a large number of very small resistive elements each shunted by a capacitance. The resistive elements result from the photoconductive material (lead oxide in modern tubes), while the capacitive partners are

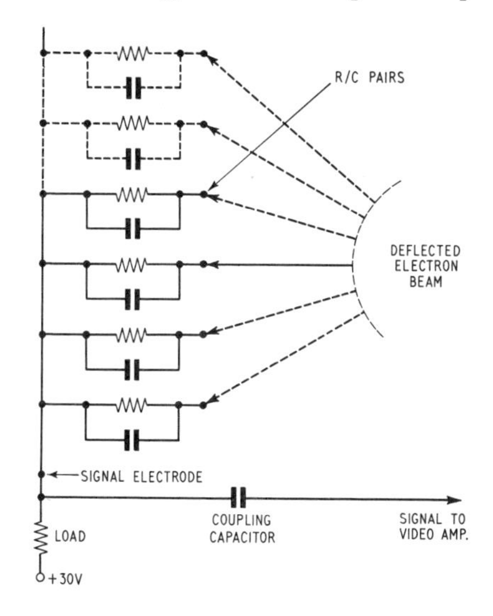

Figure 3.4. Equivalent circuit of photoconductive camera tube, showing the effective RC elements (see text)

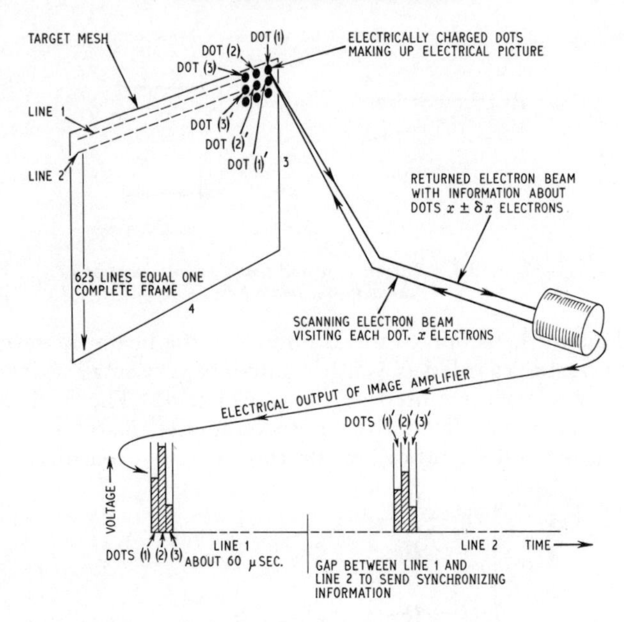

Figure 3.5. Basic principle of scanning at the camera tube

created by the photoconductive material itself as one plate and the transparent conductive film, shown in *Figure 3.3*, as the other plate. One side of the combinations is common to the target electrode, while the other side of each RC pair is effectively open-circuit until the electron beam comes into play.

As the beam 'point' scans the target area each RC pair is connected in turn for a fleeting moment to the beam itself, and since the beam is composed of electrons it is the equivalent of a conductor carrying an electric current. The beam thus constitutes a conductor of diminutive momentum which can easily be deflected electrically at high rates.

The equivalent circuit of the target area in terms of RC elements is given in *Figure 3.4*. Now, when the beam falls in turn upon the elements, each one picks up an electric charge, the charging circuit being from the positive side of the load resistor connected to the target or signal electrode as it is called, through the load resistor, across the capacitance, through the electron beam and gun and back to the negative side of the signal electrode potential source.

Because the capacitance of each element is shunted by a resistance the charge picked up from the beam as it scans commences to drain away as soon as the beam passes the element. This action is best contemplated relative to a single element. As soon as the beam passes this the charge on the capacitance starts to flow out through the resistance, and by the time the beam comes round on the next scan the capacitance will have lost a certain value of charge which, to be restored, results in a flow of current through the load resistor.

Assume that the camera is operated in the dark so that there is no light falling on the tube target area. The photoconductive target will thus have a very high resistance, meaning that very little or no charge will be lost from the capacitance of each element, so no current flows through the load resistor since there is no charge loss to compensate for.

Conversely, when the camera is responding to an image the photoconductive material will have a much lower resistance at each element upon which light is falling, so a relatively large current will flow through the load resistor as the beam scans these elements.

The signal representing the various light values of the image thus consists of the voltage developed across the load resistor resulting from the changing current through it as the beam scans the image focused on the target area. Recent cameras employing lead oxide tubes have very high sensitivity, some giving pictures down to 50 lux (5-foot candles), while working normally at 500 lux (50-foot-candles). Tube diameter is around 30 mm, so one can appreciate how much smaller is the image falling on the sensitised layer of the camera tube than that on the screen of the display tube.

Sync pulse generator

The three beams of a three-tube colour camera must scan accurately together, and for a picture to resolve properly at the receiver the vertical and horizontal beam deflections at the camera must be synchronised with those at the receiver.

Synchronising is provided by pulses applied to the picture signals at the intervals between the scans. They originate from a generator which is under the control of a master oscillator.

The master oscillator signal, which is controlled very accurately, is of a frequency of several megahertz (MHz), and is counted down to the required line and field frequencies by dividing circuits and special gating and mixing circuits. Circuits are also incorporated to introduce the black level intervals and the colour bursts, these being looked at again in subsequent chapters.

Timebases

The electron beam of each camera tube scans one line of the target image in just over sixty millionths of a second (60 microseconds, μs), and then returns at the end of the line to the beginning again in about 10 μs. At the same time it is deflected downwards less fast and, as we have seen, at the end of each field returns to the top of the image in a shorter time.

These deflections require the application of specially shaped waveforms to the deflecting mechanism of the camera tubes. The waveforms are rather like a sawtooth in shape, and for this reason are known by this name. The bottom part of *Figure 3.6* illustrates a sawtooth wave. The same sort of waveform is used for both the horizontal and vertical scans, the only difference being in the time scales, for the sawtooth for field deflection has a time period several hundred times as great as that for line deflection. This idea is brought out in *Figure 3.7*. Both of these waveforms are

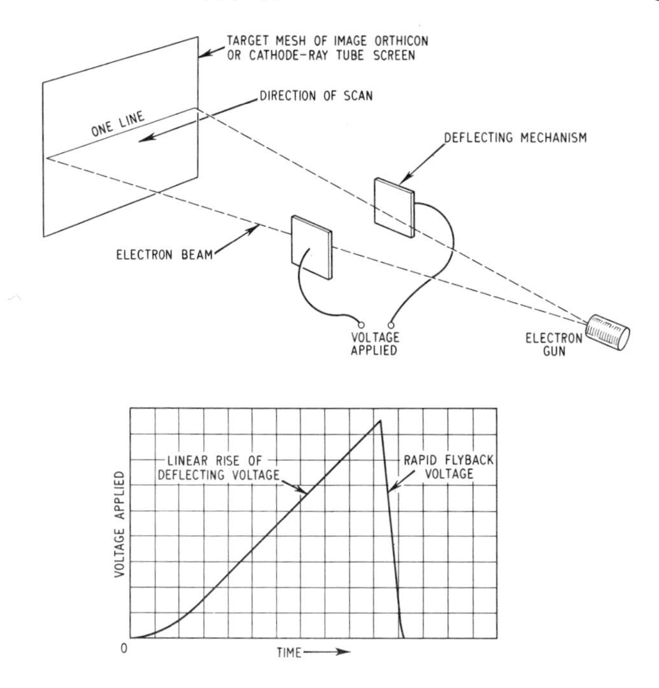

Figure 3.6. The top drawing shows how an electron beam can be deflected electrostatically. The deflecting plates receive a sawtooth voltage waveform as shown in the bottom drawing. Because the beam is composed of electrons, which are negative charges, the beam is attracted to the plate rising positively. A 'push-pull' arrangement is commonly used whereby while one plate rises positively and attracts the beam, the other rises negatively and repels it, this adding to the deflecting force. An electron beam can also be deflected magnetically since a beam of electrons can be regarded as a current flow. For this sawtooth current waveforms are applied to coils (see the vidicon tube in Figure 3.2)

produced by circuits called timebases, the *line timebase* for the horizontal scan and the *field timebase* for the vertical scan. Each camera requires these circuits as also does each television receiver. Their job is to keep the scanning beams of the camera tubes and the scanning beams of the receiving picture tubes working.

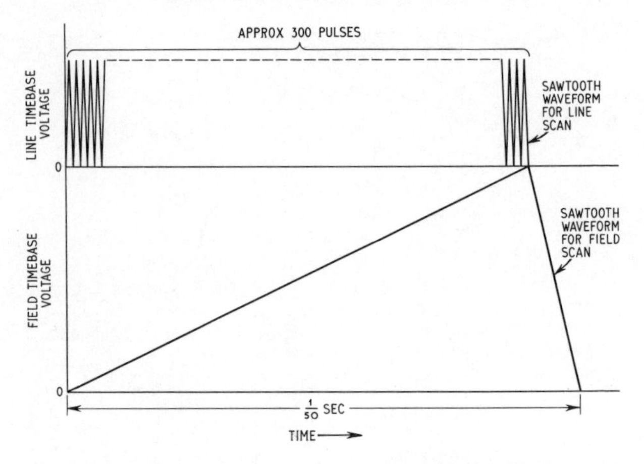

Figure 3.7. Comparison between the time scales of line and field scanning

Synchronisation

The sync generator provides two sets of sync pulses for the camera tubes and two sets of identical pulses for the television receivers—the two sets, of course, are the line and field pulses. There is no trouble in providing the camera tubes with the pulses, for all that is required is a short length of screened cable between the generator and the camera. But there is more fun in getting them to the homes of the viewers, and the only way that this can be done is to introduce them to the picture signals at the studio prior to main transmission. The sync pulses can be regarded as 'timing' the scan functions of the timebases in rather the same way as a balance wheel 'times' a watch.

The sync pulses to the camera tubes are supplied a little in advance of those introduced to the signal. This is because the camera tubes must be working and sending out picture information for a brief moment before the picture tube of the receiver starts to scan.

The functions of the sync generator are common to both monochrome and colour television systems.

The encoder

The three primary colour signals from the three camera tubes are communicated to the input of the encoder, and this processes them ready for transmission. The process can be likened to the sending of an important letter in code. If a code book is used at the dispatch end the letter can be reduced to a shortened form and sent to the recipient who, using a similar code book, can decode it and thereby get it to appear in its original form.

The three primary colour signals are first added together in the proportions required for the luminance or Y signal, as explained in Chapter 2. The Y signal is then separately subtracted from the red and blue primary colour signals to give the two colour-difference signals (R−Y and B−Y). After weighting (as in the PAL system for example), the two signals are amplitude modulated in quadrature upon a subcarrier which is subsequently suppressed so that only the sidebands of the R−Y and B−Y signals remain. It will be recalled that these are the V and U chroma signals in the PAL system.

With monochrome television (see, for example, *Beginner's Guide to Television*) only the luminance (known as video) signal and the sync pulses with the various black level intervals are attached to the carrier wave by the process of amplitude modulation; that is, the amplitude of the carrier wave is varied by the video signal and sync pulses, etc. The information is then propagated over the area served by the transmitter for reception by individual viewers.

With the need to send a chroma signal in addition to the luminance signal, coupled with the need to keep the composite signal within the channel bandwidth normally occupied by a monochrome signal, the method of fitting together all the components of a signal for compatible colour television is necessarily more complex. As already intimated, the scheme utilises a subcarrier of a much lower frequency than the main carrier wave. The latter, in fact, may be several hundred MHz, in television Bands IV and V, while the former is a few MHz only, the actual frequency being geared to the line timebase repetition frequency.

With the PAL and NTSC systems the subcarrier is quadrature amplitude modulated by the R—Y and B—Y signals, and when the subcarrier is suppressed the result is the chroma signal, as already noted. The chroma signal is combined with the luminance signal, sync pulses and intervals, etc., the integration then forming the composite colour television signal, and it is this which amplitude modulates the main carrier wave.

We shall be seeing later that to get the chroma signal to integrate neatly with the luminance signal the frequency of the subcarrier must be related to the repetition frequency of the line timebase. The extra chroma information is then 'squeezed' into discrete intervals of low energy between the luminance sidebands in the overall spectrum. It is by this technique that the extra chroma information can be carried in a standard 625-line television channel with the least interference, especially to black-and-white pictures produced by monochrome receivers working from a colour signal.

These rather special techniques, of course, tend to enlarge the task of the sync pulse generator, for its master oscillator must be correctly related to the frequency of the subcarrier generator. However, in practice the master oscillator is generally used as the subcarrier generator, and this automatically takes care of the relationship between the line timebase repetition frequency and the subcarrier frequency.

To summarise, therefore, the colour-difference signals are used to amplitude modulate a subcarrier in quadrature, after which the subcarrier is suppressed so that the sidebands only of the colouring information appear when colour is in the scene. This is the chroma signal which is added to the luminance signal in a manner relating to the line timebase

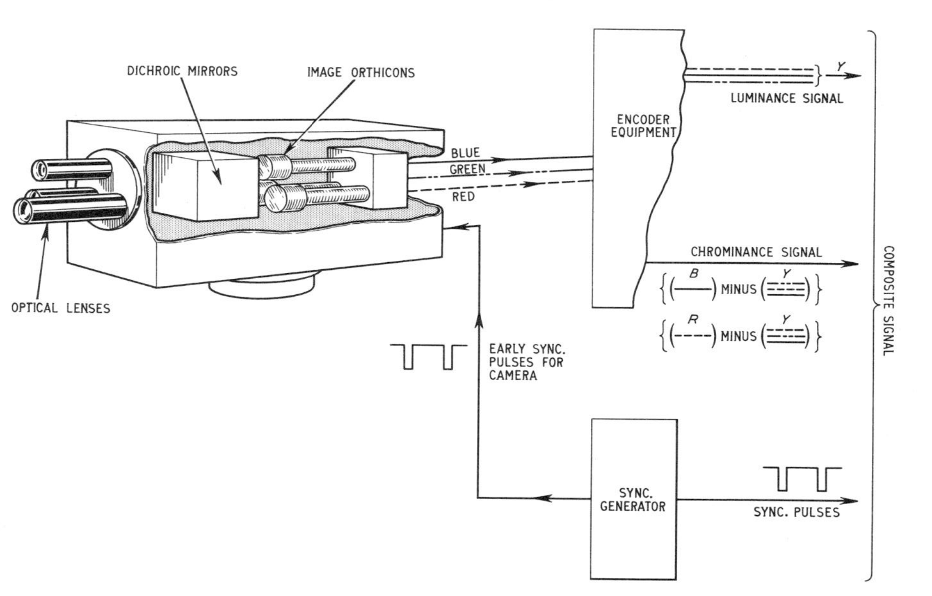

Figure 3.8. *Showing in basic terms how the composite signal is developed. Note that the chroma signal is produced by the $B-Y$ and $R-Y$ signals amplitude modulating a subcarrier in a special way, called quadrature modulation. The subcarrier is suppressed so that only the sidebands of the colour-difference signals remain. At the receiver detection is possible only by reintroducing the subcarrier*

frequency and subcarrier frequency. Both are finally combined with sync pulses, intervals and colour bursts to form the composite signal which goes to the transmitting station where it is used to amplitude modulate the high power u.h.f. carrier wave.

This description applies both to the NTSC and PAL system. The SECAM system differs in the way that the chroma signal is prepared and attached to the subcarrier, and is treated more fully in Chapter 10.

4

THE COMPLETE SIGNAL

The previous chapters have outlined the nature of the signals required for a colour television system. However, to understand how the transmitter and receiver, particularly the latter, work it is necessary to know a little more about the complete signal, how it is formed and how the PAL signal differs from the NTSC signal from which it was derived.

First let us recap a little. We have seen that the luminance and chroma signals, which are transmitted simultaneously, provide a compatible system; that the luminance signal is equivalent to the ordinary video signal of a black-and-white system; and that the luminance signal is produced by combining the three primary colour signals in specific proportions to satisfy the physiology of the eye, though in some cases it might be produced by a fourth tube in the colour television camera.

Video/luminance signal

Although this is a book concerned essentially with colour television, we really need to know a little more about the make-up of the luminance signal before we can fully understand how the chroma signal is added to it. It will be recalled that on each line scan the electron beam in the camera samples the brightness levels of the image along discrete lines, the whole image area being so explored by the beam moving relatively slowly downwards as it is moving from left to right. Depending on the brightness level of the image at any point, the camera tube gives an output of voltage which, from the luminance tube of a colour camera or from the solitary tube of a monochrome camera, can be regarded as the *video signal*. If the line of the image which is being

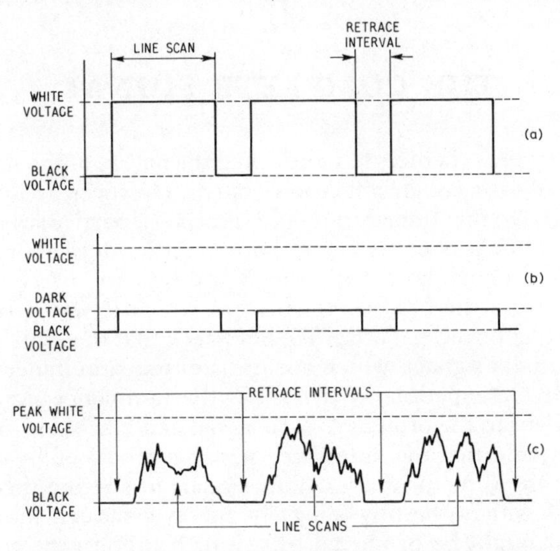

Figure 4.1. Lines of video signal. (a) *high brightness.* (b) *low brightness.* (c) *picture content*

scanned is of a constant high brightness level, the video signal will be of a relatively high value and constant, as shown at (a) in *Figure 4.1*. If the brightness is low but constant, then a signal similar to that at (b) will result. However, when the brightness along a scanned line varies, as it usually does in a picture, a signal something like that shown at (c) will result.

This is the video signal of monochrome television and the luminance signal of colour television. Remember that the luminance signal from a colour camera can be produced by the correct addition of red, green and blue primary colour signals from the three camera tubes when there is not a separate tube for the luminance signal.

The periods between the line scans allow the beam to return from the left- to the right-hand side of the image to start the next scan, and here the line sync pulses are located. The best way of showing them is by a real video signal picture, taken from the screen of an oscilloscope. Such a picture (called oscillogram) is given in *Figure 4.2*, and this

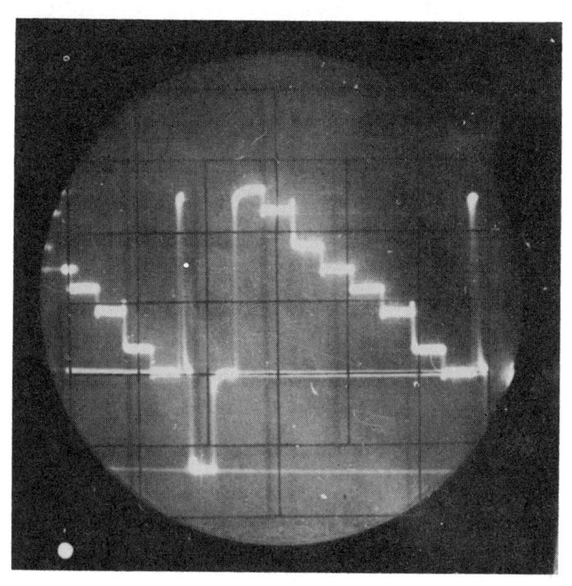

Figure 4.2. One and half lines of video signal resulting from the luminance gradations of the colour bars. The steps correspond to the different brightness levels of the bars. The line sync pulse and porches are also shown between the lines of video signal

shows about one and a half lines of video signal resulting from the luminance values of the colour bars. The bars start at white and then gradually fall in luminance value to black. Hence the staircase effect of the video on the oscillogram.

Colour bursts

In the interval between the first half line shown in the oscillogram and the next full line is the line sync pulse which, videowise, goes below black level, into the region colloquially termed 'blacker-than-black'. To permit stabilisation of the video and sync circuits after a line of video and a sync pulse,

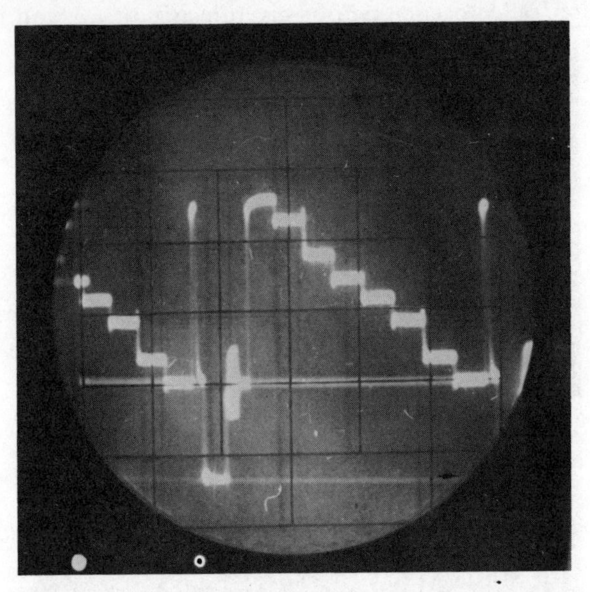

Figure 4.3. This oscillogram is similar to that in Figure 4.2, but it also shows the colour bursts generated on the back porch to the line sync pulse

particularly in the receiver, very short duration intervals at black level occur either side of each line sync pulse, after one line of video signal and before the next line. With colour signals the bursts are generated on the second interval, as shown by the oscillogram in *Figure 4.3*. This is virtually the same as that in *Figure 4.2*, but with a colour burst applied (also see *Figure 2.7*). The starting interval is sometimes called the front porch to the line sync pulse and the finishing one the back porch, and it is on the latter that the colour burst is generated.

On a monochrome signal, of course, one would not normally find a colour burst; but the oscillogram in *Figure 4.3* was specially prepared to show merely the luminance information of the colour bars (as in *Figure 4.2*) along with a colour burst. When there is a colour burst, of course, one would also normally expect to see chroma signal, but we shall be looking at this in a minute.

Interlaced scanning

Broadcast television systems adopt so-called interlaced scanning to reduce the display flicker, and as this is common to both monochrome and colour television this is not really the book in which to introduce this particular principle (see, for example, *Beginner's Guide to Television*). Nevertheless, a basic conception of the idea is desirable since it can influence certain display aspects of PAL colour television.

A complete picture, which is sometimes referred to as a *frame*, is composed of two *fields*, with the lines of one field occurring between the spaces of an adjacent field. Thus, if a television picture is made up of 625 lines, then one field will contain 312½ lines, so that the interlaced lines of the two fields together will constitute a complete picture of 625 lines. The pulses for field synchronising are timed to ensure that the field scans are initiated for correct interlacing.

Composition of the chroma signal

So far, then, we have seen how the luminance signal is formed line-by-line and how the line sync pulses and colour bursts are introduced. We must now achieve a better understanding of the chroma signal and see how this is added to the luminance signal, etc. to form the composite signal.

It has already been stated several times in previous chapters that the R−Y and B−Y colour-difference signals yielded by the camera signal matrix are amplitude modulated in quadrature onto a subcarrier. Just what does this mean? We probably already know something about amplitude modulation to the extent that the modulation signal of lowish

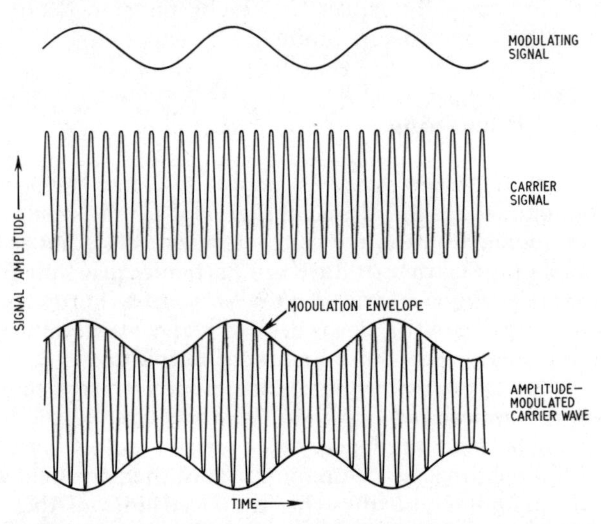

Figure 4.4. The amplitude modulated waveform is produced by the carrier signal being modulated with the modulating signal. Note that the waveform inside the modulation envelope results from the addition of the carrier wave (f_c), the upper sideband (f_c+f_m) and the lower sideband (f_c-f_m), where f_m is the modulation frequency

frequency modulates the carrier wave of higher frequency by causing its amplitude to vary in concord with the modulation signal. *Figure 4.4* illustrates the effect. Here a single sine wave signal is shown modulating a higher frequency carrier wave. This modulation is not all that difficult to understand from an elementary viewpoint, and it can be achieved by the use of simple circuitry. For example, the modulating signal can be applied to a valve or transistor so that it effectively alters the supply voltage on the anode or collector. Thus when the carrier wave is applied to the grid or base the amplitude of the output signal alters to the pattern of the modulation waveform. When a pure sine wave is the modulation signal, therefore, the carrier wave amplitude varies in a like pattern and the carrier wave is then said to possess a modulation envelope. This is shown in *Figure 4.4*.

Any type of modulation gives rise to sideband signals, and with amplitude modulation there is an upper and a lower sideband for each modulation frequency. If the carrier wave is, say, 10 kHz and the modulation frequency a pure 1 kHz sine wave (e.g., a single modulation frequency), then the upper sideband will be $10+1 = 11$ kHz, while the lower sideband will be $10-1 = 9$ kHz. This simple arithmetic follows for all single modulation frequencies. Thus the modulator delivers three signals. The carrier wave, the upper sideband and the lower sideband. The information of the modulation is effectively present in the sidebands, so it is possible to suppress the carrier after modulation for transmission, though for demodulation at the receiver the carrier wave will have to be reintroduced very accurately somehow.

So much, then, for the amplitude modulation of one set of information; but what about the modulation of two sets of information which are the R−Y and B−Y colour-difference signals? This is where the quadrature amplitude modulation of the subcarrier comes in. In the British system of colour television the subcarrier frequency is accurately controlled at 4.433 618 75 MHz. This is usually referred to roughly as 4.43 MHz. The numerous decimal places are necessary for various reasons, one of which is to minimise interference of 'dot-pattern' type which can arise on a received picture due to the relationship between the subcarrier frequency and the line timebase frequency. The frequency accuracy is required to secure the optimum relationship in this respect, and hence the least dot-pattern interference. It is also necessary to have the subcarrier within the video bandwidth, and as high in frequency as possible for minimum interference. In the British system both the upper and lower sidebands are exploited equally, and this means that the subcarrier must be placed in frequency so that both sidebands can be fully accommodated in the video spectrum.

The subcarrier is obtained in the correct relationship with the line timebase frequency from the sync generator master oscillator as we have already seen. The oscillator drives *two* amplifiers which produce two subcarrier outputs, but with one having a 90-degree phase difference with respect to the other.

Now, for the benefit of readers who may be uncertain as to what this means the following explanation may be helpful. A complete cycle of alternating current or voltage can be regarded as occupying 360 degrees—a circle! Although this statement is somewhat arbitrary it is supported by sound reasoning. Consider a generator, for example; this will, in the simplest case, rotate through 360 degrees and during that period produce a complete sine wave. If a second generator is set going but a quarter turn ahead of the first, then the sine wave produced by this will be 90 degrees out of phase with that yielded by the other one.

Since in the colour television case the one oscillator drives two amplifiers the frequency of the two outputs will be absolutely identical and, moreover, the synchronism will be maintained, but always with one output 90 degrees ahead of the other owing to the deliberately arranged 90-degree phase shift provided by circuit elements.

Figure 4.5 highlights the situation, where the two full-line sine waves are of exactly the same frequency but 90 degrees out of phase. The x axis is calibrated from 0 to 360 degrees corresponding to a complete cycle of the first full-line sine wave. The second one starts a little later in time; in fact, when the first has arrived at the 90-degree mark. Thus we have a direct illustration of the 90-degree phase difference between the two subcarriers which, remember, are derived at the transmitter from a common oscillator or generator.

One of them is amplitude modulated with the R—Y signal and the other with the B—Y signal. After suitable weighting, therefore, they yield the V and U chroma signals of the PAL system or the I and Q signals of the NTSC system (see

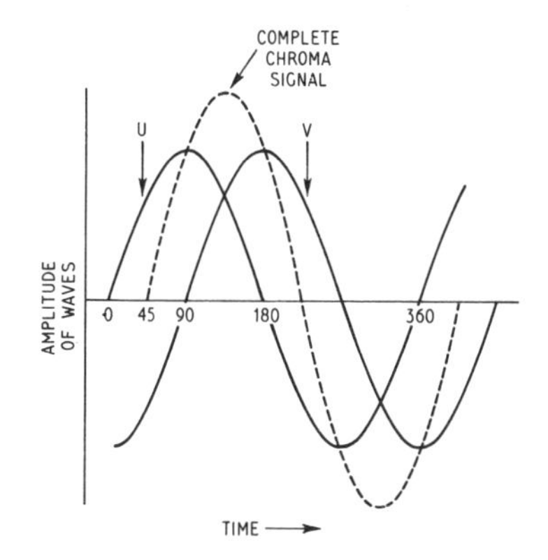

Figure 4.5. The V and U waveforms have the same frequency and amplitude but differ in phase by 90 degrees. The broken-line waveform represents the chroma signal complete which is the quadrature addition of the V and U signals

Chapter 2). Since this book is primarily concerned with the PAL system let us label one V and the other U.

The next move in the game is to add the V and U signals to form the chroma signal proper, and this signal is shown by the broken-line sine wave in *Figure 4.5*. The fundamental feature of this mode of signal addition is that by special detection at the receiving end it becomes possible to isolate the V and U signals again and hence extract the original

R−Y and B−Y modulation signals. This is possible, of course, because of the fixed 90-degree phase difference between the two signals. Signals of the same frequency and phase lose their individual identity for all time when added.

In colour television literature the various colouring signals are regarded in terms of vectors. These are mathematical devices for displaying the specific features of amplitude and phase of a signal at one instant in time. The basic vector 'background' is given in *Figure 4.6*, where our complete circle of 'signal' is represented by the four quadrants. Motion or time is regarded as anticlockwise, so starting from the zero-degree datum we follow the angles of the signal as shown by the arrowed circle. Colour signals are generally more complicated than implied by simple vectors and phasors, so this method of presentation may fail to illustrate a situation fully, and should thus be viewed accordingly.

Developing this theme in terms of the V and U signals in *Figure 4.5* we get the vector diagram of *Figure 4.7*. The arrowed lines here correspond to the amplitudes of the three signals—the V and U signals and the complete chroma signal. The amplitude of the complete chroma signal is obtained by completing the square (it could, of course, be a parallelogram, and is usually referred to as this, when the amplitudes of the V and U signals are unequal) as shown. This diagram clearly shows that the V and U signals have a 90-degree phase difference and that the amplitudes of the two signals in this case are equal. It also shows that the phase of the complete chroma signal falls half-way between the zero-degree datum and 90 degrees, which of course is 45 degrees. Note clearly, though, that the phasing angle of the complete chroma signal is only at 45 degrees when the V and U amplitudes are equal. The angle changes within the quadrant when there is a relative difference between the amplitudes of the V and U signals. Should the amplitudes alter together, then the original angle holds. That the complete chroma signal is 45 degrees relative to either the U or V signal is proved in *Figure 4.5*. However, it is much more

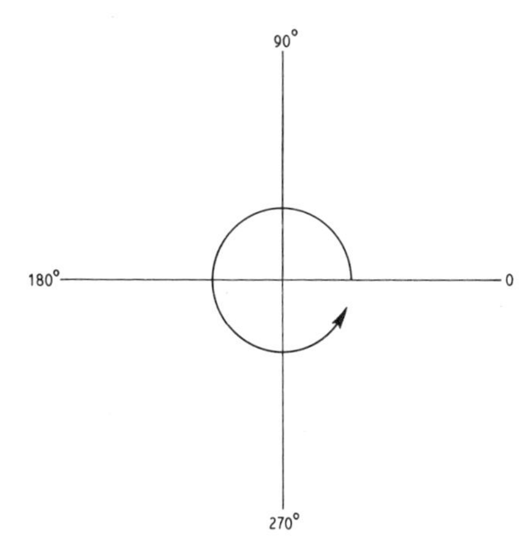

Figure 4.6. Vector 'background' diagram in four quadrants. Time or motion is regarded by convention as being anticlockwise from the 0-degree datum, as shown by the arrowheaded circle

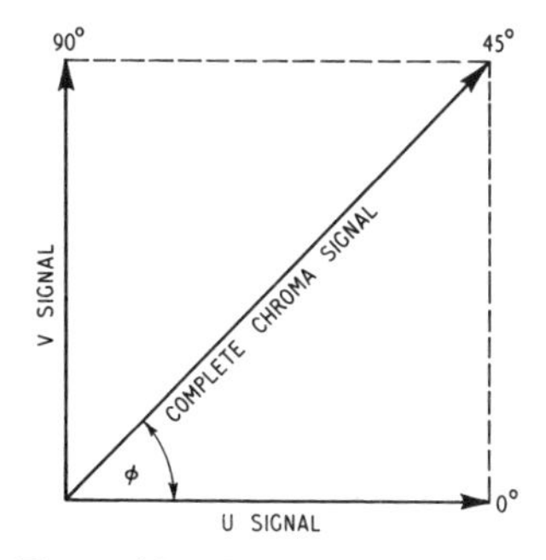

Figure 4.7. Vector representation of the V, U and complete chroma signals of Figure 4.5 (see text)

convenient to work with vectors than with complex waveform diagrams as will be appreciated.

We can easily discover the amplitude of the complete chroma signal which, in colour television at least, is often referred to us the phasor, from

$$\text{Phasor amplitude} = \sqrt{(V^2+U^2)}$$

Thus when the V and U signals have equal amplitude, the phasor amplitude is 1.4 times that of V or U. Readers with some knowledge of trigonometry will also see that the angle $\varnothing$ of the phasor in the quadrant of *Figure 4.5* is equal to $\tan^{-1} V/U$.

More about sidebands

Let us leave this basic theory for a minute and return to the transmitter. So far then we have a very happy state of affairs: all the luminance information is in one signal, while the colouring information is neatly packed away into a single subcarrier. Since the subcarrier has a relationship to the luminance signal (e.g., by it being a multiple of the line scanning frequency) its sidebands interleave in the gaps left by the sidebands of the luminance signal, so that no extra band space is required for the colouring information.

At this juncture a few more words about sidebands would not be amiss. We have seen that when a carrier wave is modulated upper and lower sidebands occur for each modulation frequency, the upper one being equal in frequency to the carrier *plus* the modulation frequency and the lower one to the carrier *minus* the modulation frequency (see *Figure 4.4*). This is for *amplitude* modulation. Frequency modulation, which is used for the sound signals and for the chroma signal of SECAM, also yields sidebands, but as a series for each modulation frequency of diminishing amplitude either side of the carrier frequency, depending on the modulation depth and frequency. The amplitude of the sidebands arising from amplitude modulation is also governed by the depth of the modulation, with 100 per cent modulation producing upper and lower sidebands each equal to half the amplitude of the carrier wave.

It will be understood, of course, that the carrier wave *carries* the information over a distance, just as an electric conductor can carry information from point A to point B. With 'radio' there is no electrical circuit, but instead the modulation (intelligence) is caused to vary the amplitude, frequency or phase of the carrier wave. The two which mostly interest us are amplitude and frequency modulation (phase modulation, in fact, is very similar to frequency modulation, but the distinction is far too complex for us to explore in a book of this relatively basic nature). With frequency modulation the frequency of the carrier is varied according to the intelligence contained in the modulation while the amplitude is kept constant, and with amplitude modulation it is the amplitude of the carrier which is varied according to the intelligence while the frequency is kept constant.

It is the variations of the carrier characteristics which produce the sidebands. If we consider an unmodulated carrier wave as a single frequency of a given amplitude the effect of varying its amplitude or frequency is to produce a number of other side frequencies of smaller amplitude. In other words, a number of extra signals is produced, which are the sidebands, each with its own amplitude and frequency obtained from the modulation in the manner already discussed.

It is noteworthy that the upper and lower sidebands of a double sideband system are mirror images of each other, so it is possible to suppress a part or the whole of one set of sidebands and yet still send the intelligence. This is, in fact, done in television to conserve channel space. That is, the lower sideband of the complete vision signal is partially suppressed giving a vestigial lower sideband of 1.25 MHz as distinct from the upper sideband extending to some 5.5 MHz.

This does not apply, though, to the chroma signal alone, for as we have seen this is amplitude modulated on the subcarriers without sideband suppression. Nevertheless, the subcarrier is suppressed at the transmitter and only the sidebands are sent by interleaving them into the low energy gaps of the luminance signal as already noted. More is said later about the subcarrier suppression.

Figure 4.8 shows the spectrum of a UK television channel, where the bandwidth is referred to the vision carrier. Notice the lower sideband suppression and the ± 1 MHz space occupied by the chroma signal.

This can be related to the sideband structure of the various components of the signal within the channel shown in *Figure 4.9*, which reveals diagramatically how the chroma sidebands are interleaved between the luminance sidebands, and also the sidebands resulting from the frequency modulated sound signal accompanying the vision signal within the channel.

Suppression of subcarrier

Because the colouring information is present in the sidebands of the subcarrier it is possible to suppress this at the transmitter and by so doing to minimise significantly the presence of dot-pattern interference on the screens of monochrome receivers using the colour signals. It will be recalled that the waveform inside the modulation envelope such as shown in *Figure 4.4* results from the addition of the carrier wave and the upper and lower sidebands. As would be expected, a modulation waveform devoid of the carrier wave differs significantly from that with the carrier wave intact since the former is composed only of the upper and lower sidebands. The diagram in *Figure 4.10* attempts in one way to reveal the salient features.

Compared with the bottom diagram in *Figure 4.4*, it will be seen that the envelope has effectively 'collapsed', so that the top and bottom parts intertwine, the sine wave in heavy

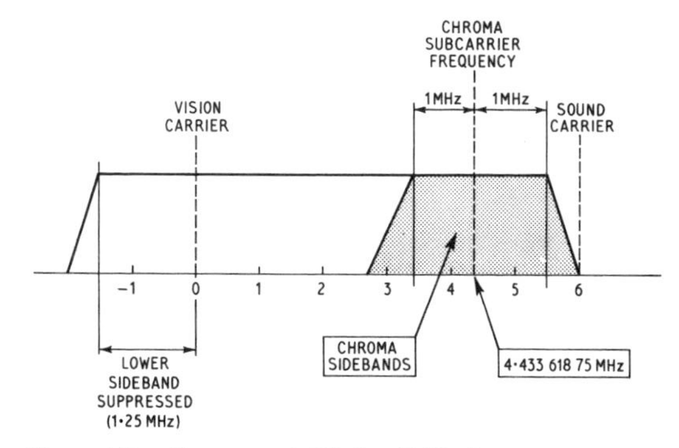

Figure 4.8. Spectrum of 625-line PAL signal in U.K. television channel

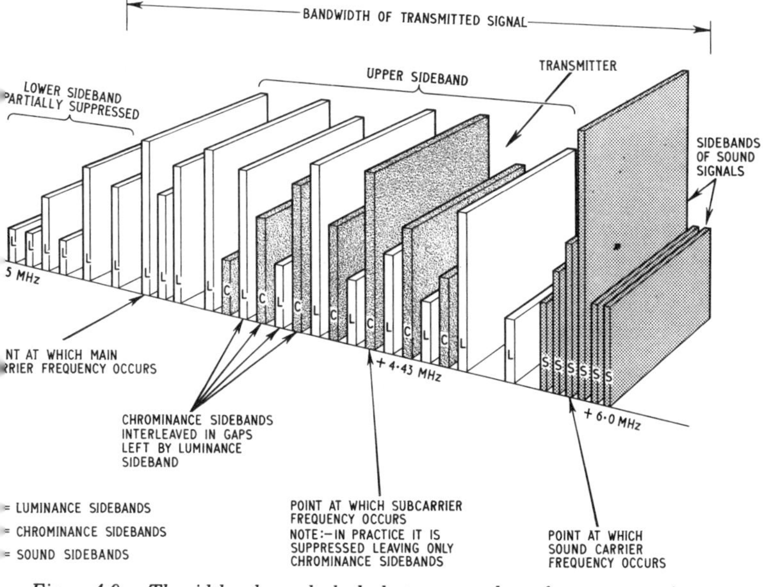

Figure 4.9. The sidebands can be looked upon as packets of energy at various frequencies

line representing the top of the original and that in light line representing the bottom of the original. Further, the high-frequency signal inside the collapsed envelope has also changed character. The frequency, however, is just the same as the original carrier wave because it is composed of the upper and lower sidebands, the average of which is the carrier frequency; but it can be regarded that phase

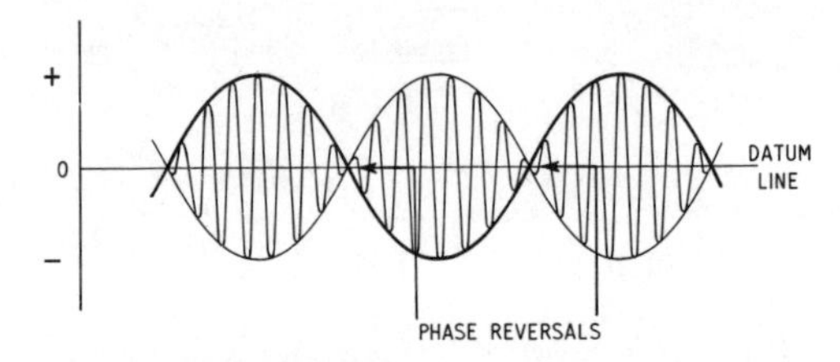

Figure 4.10. Modulation waveform when the carrier wave is suppressed. Compare this with the waveform at the bottom of Figure 4.4. An important aspect is the phase reversal of the high-frequency wave each time the modulation sine waves cross the datum line. This waveform is the addition of the upper and lower sidebands only

reversals occur each time the sine waves representing the top and bottom parts of the original envelope pass through the datum line. It is difficult to show these diagramatically; but the high-frequency signal changes phase by 180 degrees at each 'envelope crossover' point, and this has a vital significance in colour television, as we shall see.

The subcarrier modulation constitutes the B—Y and R—Y signals, which change continuously during a programme. However, during a test transmission, say, of the colour bars, there is much less change of information during a line scan, for the colour signals remain constant over each bar per line scan, changing only from bar to bar.

The diagram in *Figure 4.11* shows the weighted B—Y signal produced by the yellow and cyan colour bars at (a), the U chroma signal modulation waveform due to the bars at

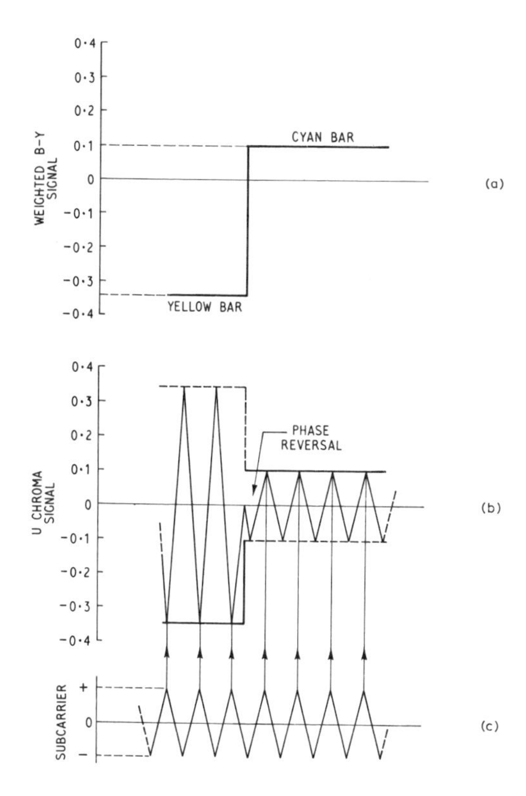

Figure 4.11. PAL weighted B—Y signal at the yellow/cyan bar transition (a). *The waveform resulting from suppressed carrier modulation* (b). *Subcarrier signal in correct phase* (c). *At the receiver the reintroduced subcarrier samples the peaks of the modulation sideband components to give an output corresponding to the amplitude and polarity of the original colour-difference signal. The polarity is given because the phase of the sideband components reverse each time the colour-difference modulation signal crosses the zero datum line (also see Figure 4.10)*

(b) and the subcarrier signal at (c). The U chroma modulation signal at (b) is with the subcarrier suppressed, and since the modulation signal is a stepped waveform going from −0.33 corresponding to the yellow bar to +0.1 corresponding to the cyan bar through the zero datum, it follows that the polarity change between the two bars will cause a 180-degree phase reversal of the high-frequency signal just the same as when the modulation is a sine wave signal, shown in *Figure 4.10*.

The phase reversal is indicated, and more clearly shown in *Figure 4.11* than in *Figure 4.10*. For example, it will be seen that the positive tips of the subcarrier correspond to the negative tips of the U chroma signal during the yellow bar and to the positive tips (owing to the phase reversal) during the cyan bar. Thus, when the subcarrier is caused accurately to sample the peaks of the chroma signal like this, information is obtained on both the amplitude *and* the polarity of the colour-difference signal. This is, in fact, how the receiver demodulates the chroma signals, the subcarrier being provided by a reference generator which is synchronised by the colour bursts.

Basic NTSC transmission system

The information so far given applies essentially to both the NTSC system and to the PAL system; but the PAL has other features which are considered in Chapter 5. At this juncture it would be desirable to collect our thoughts and to have a look at the basic NTSC transmission system (see Chapter 5 for that of the PAL system). The block diagram of this is given in *Figure 4.12*. Notice here that we have Q and I modulators instead of the V and U modulators of the PAL system.

Important parts of the composite signal are the luminance and chroma signals, the sync pulses with intervals and the colour bursts.

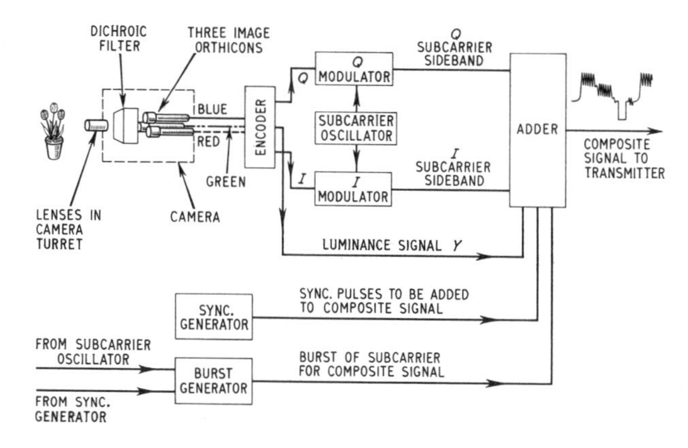

Figure 4.12. Block diagram illustrating the basic NTSC transmission system

The process of adding these parts together can be understood by considering each one as being produced by a separate generator, with the generators then being connected in series with a load resistor, across which the composite signal occurs. The idea is shown in *Figure 4.13*.

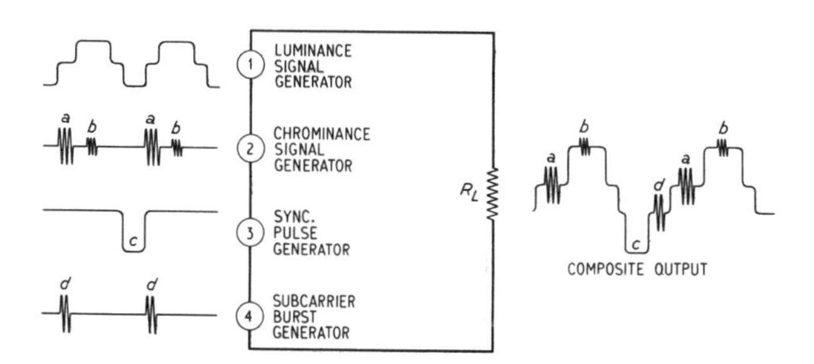

Figure 4.13. The addition of the various components of the composite signal can be represented by considering the output across a load resistor R_L whose current is derived from a number of generators connected in series

When the composite signal leaves the studio its form is roughly after the style of that shown in *Figure 4.14*; but here the time intervals and amplitudes of the various components are not to scale. Beginning at point A on the diagram is the scanning of one line on the camera tubes. The line from A to D represents the luminance signal, which is the combined output of the three tubes or that from the fourth 'luminance' tube during one line scan. It represents the brightness of the picture and is thus very similar to the signal produced by a monochrome system. The high-frequency alternating signal superimposed on the luminance signal consists of the chroma sidebands which, in the PAL system, have a frequency close to 4.43 MHz, as we have seen. These are really of since wave

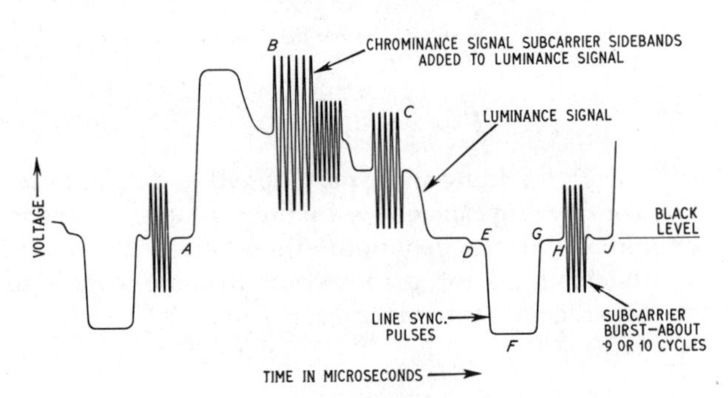

Figure 4.14. An impression of the composite colour television signal (not to scale)

nature and not triangular waves as shown. Similarly, the triangular waves drawn in *Figure 4.11*, which again consists of the chroma sidebands, are meant to represent sine waves. It is easier to draw triangular-type waves than sine waves and to reveal a phase change with them, as in *Figure 4.11*. The NTSC subcarrier frequency on the American 525-line system is 3.595 45 MHz. When the NTSC system was being experimented with by the BBC during 1962–63 on 625 lines the frequency was 4 429.687 5 MHz.

Now back to *Figure 4.14*. From D to E is a pre-sync line-suppression interval, where the brightness component falls to a voltage which corresponds to illumination threshold on the display screen, so that events on the signal at this level or below (e.g., blacker-than-black) fail to be observed by the viewer. The camera at the studio is also blanked off during this period and no picture information is transmitted.

The camera and display tube electron beams are at this time 'retracing' prior to starting another line scan, such action being instigated by the line sync pulse EFG. This ensures that the beams of the display tubes are always synchronised to those of the camera tubes.

Following the sync pulse is another black level interval, called the post-sync line suppression, G to H in *Figure 4.14*, and this is followed by the colour burst of 9 ± 1 cycles of subcarrier frequency from H to J. The Americans called the two black level intervals front and back porches to the line sync pulses, and these names are still sometimes used today in the U.K. The back porch has a longer duration than the front porch, and it is on the former that the colour burst is generated as we have seen. The first is inserted to give the receiver circuits time to return to black level before the sync pulse commences, even when the extreme edge of the picture is of high illumination, while the second puts a black border on the left-hand side of the picture, which provides a brightness level reference.

The chroma information from B to C extends beyond peak white on the luminance information, but weighting is employed to prevent the main carrier wave being overmodulated by the composite signal. The 625-line system adopts so-called negative modulation, opposite to the positive modulation of the old 405-line system, which means that maximum modulation depth occurs at the blacker-than-black sync pulse tips, while peak white corresponds to almost zero modulation depth.

The composite signal is conveyed to the transmitter where it is used to amplitude modulate the main carrier wave corresponding to the appropriate ultra high-frequency (u.h.f.) channel in Band IV or V. *Figure 4.15* (left-hand side) shows the modulated carrier with negative modulation, where the line sync pulses are on the outer envelope corresponding to maximum modulation depth, and for comparison the old 405-line monochrome positive modulation, where the sync pulses are at low modulation depth inside the envelope.

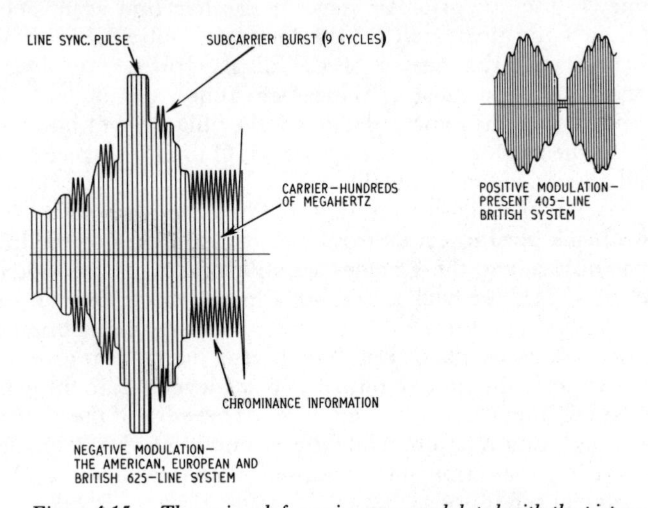

Figure 4.15. The main u.h.f. carrier wave modulated with the picture signal. Left, negative modulation due to the composite colour signal and, right, positive modulation due to early 405-line monochrome signal

Of course we have only examined one line of signal. In practice there are 625 of them, though not all of them carry picture information because a few are blacked out at the end of a field and the start of the next to accommodate the field sync pulses and associated black level intervals.

Sound signal

Accompanying the carrier-wave-modulated composite signal is the sound signal which uses a separate carrier wave and adopts frequency modulation. The position of this is shown in *Figure 4.8* and the sidebands associated with it in *Figure 4.9*. It will be seen that in the British 625-line system the sound carrier is located 6 MHz away from the vision carrier. The beat between the two carriers is exploited in the receiver in terms of so-called intercarrier sound, where the 6-MHz beat signal, carrying the f.m. sound, is amplified and fed to an f.m. detector which extracts the modulation signal and directs it to the loudspeaker. More is said about this in Chapter 9.

We have now acquired an overall picture of the composite colour signal. To some extent this is common to both the NTSC and PAL systems. However, the PAL signal, which is a development of the NTSC signal, is endowed with several extra features as a means of 'locking' the hue at the receiver to that actually transmitted. With the NTSC system the hue (e.g., colour) can drift with changes in phase of the chroma signal in the system as a whole, including the transmitter, the ether through which the signal is transmitted, and the receiver. This happens because the colour of a displayed picture element is directly geared to the phase of the chroma signal—e.g., to the position of the phasor in its vector background (see *Figures 4.6* and *4.7*) at any instant. The instantaneous amplitude of the phasor determines the saturation of the elemental colour display. However, before we become too involved in this we need to know a little more about the colour-difference signals. These and the PAL system are considered in the next chapter.

5

THE PAL SIGNAL

In Chapter 2 it is shown that the colour-difference signals fall to zero when the scene is devoid of colour, the luminance signal then controlling the display in shades of grey only. This means also that the chroma signal disappears when there is no colour in the scene because the two transmitted colour-difference signals are responsible for the chroma sidebands, the subcarrier having been suppressed at the transmitter.

Figure 4.11 in Chapter 4 shows how the subcarrier is related to the colour-difference signal. Although this applies to the B−Y and U chroma signals, the principle is the same so far as the R—Y and V chroma signals are concerned. It will be recalled that at the transmitter the two lots of chroma signals—the I and Q signals of the NTSC system and the V and U signals of the PAL system—are combined in quadrature to form the complex chroma signal. In the NTSC receiver the original R−Y and B−Y signals are extracted direct from the complex chroma signal, while in one form of PAL receiver—that which is mostly in use, called the PAL-D receiver because it incorporates a chroma delay line whose purpose will become clear as we progress—the complex chroma signal is first converted back to its original V and U chroma components before the R−Y and B−Y signals are extracted. This receiver function together with other parameters of the PAL signal 'lock' the displayed hue to that in the scene irrespective of system phase distortion.

In Chapter 4 the simplified phasor concept of the chroma signal was introduced, and before proceeding we must have some idea of how this is related in terms of angle to the colour-difference signals which modulate the quadrature sub-carriers.

Clues to this are, in fact, contained in *Figure 4.11*. Here is revealed that the colour-difference signals can have positive or minus values, and that because the subcarrier is suppressed each time there is a polarity change there is a phase reversal of the high-frequency signal of which the upper and lower sidebands are composed.

More about colour-difference signals

To get to grips with all this we shall first have to cast our minds back to the primary colour signals from the camera tubes (Chapter 2). Let us suppose that the scene is red, then we get an output from the red tube only (in reality 100 per cent saturation is rarely if ever achieved, so we get a little output also from the green and blue tubes which, with the Y-proportioned output from the red tube, would be responsible for a little desaturating white, but for the sake of simplicity of description let us ignore this and assume 100 per cent saturation). If it is also assumed that each camera tube is adjusted to yield unity signal voltage at full saturation, then the R−Y signal on a red input will be +0.7 V. This is simply because the proportion of red signal in the Y signal is 30 per cent (see Chapter 2). Thus we have R−Y = 1−0.3 V = +0.7 V.

On a cyan input we get R−Y = −0.7 V. This is because the red tube yields zero output while the Y signal is 0.7 V, made up of 59 per cent green signal and 11 per cent blue signal, remembering that cyan is a light mix of blue and green. The B−Y signal on a red input is −0.3 V because the blue tube gives zero output while the Y signal is 0.3 V, made up only of the proportioned signal from the red tube.

Similarly, he B−Y signal on a cyan input is +0.3 V because the blue tube gives 1 V while the Y signal is 0.7 V composed of 59 per cent green signal and 11 per cent blue signal.

If we want to, we can easily find the values for the R−Y and B−Y signals for each of the six standard colour bars based on 100 per cent saturation and 100 per cent amplitude geared to unity output at each camera tube in the manner just illustrated. Textbooks on colour television delve into this more deeply (see, for example, the author's *Colour Television Servicing*, by the same publishers). At this juncture, however, it is worth noting that one test card of the BBC is based on 95 per cent saturation and 100 per cent amplitude, while some tests of the IBA and programme contractors are based on the EBU (European Broadcasting Union) bars which are 100 per cent saturation and 75 per cent amplitude.

What we have discovered so far, therefore, is that the colour-difference signals have both positive and negative values depending on the colour in the scene. By looking again at *Figure 4.11*, it will become apparent that the chroma signals also reflect the polarity of the colour-difference signals in terms of phase.

In the PAL system weighting is applied such that the equivalent V and U signal amplitudes are below those of the R−Y and B−Y signals. It will be remembered that this weighting is introduced to avoid overmodulation of the main u.h.f. carrier wave. The weighting values are:

$$V = 0.877(R-Y) \text{ and}$$

$$U = 0.493(B-Y).$$

Thus on a cyan input based on 100 per cent saturation and amplitude we obtain U = +0.1479 and V = −0.6139 from the B−Y = +0.3 and R−Y = −0.7 signals earlier computed. Other parameters are the amplitude and angle of the phasor, and these can also be discovered from a little geometry and trigonometry as explained in Chapter 4.

To summarise, however, *Table 5.1* gives all the parameters so far discussed relative to the hues of the standard colour bars. All these values are based on 100 per cent saturation and amplitude.

Table 5.1

Colour	Y	B—Y	R—Y	U	V	*Phasor* *Amplitude*	*Angle (deg.)*
Yellow	0·89	—0·89	+0·11	—0·4388	+0·0965	0·44	167
Cyan	0·7	+0·3	—0·7	+0·1479	—0·6139	0·63	283
Green	0·59	—0·59	—0·59	—0·2909	—0·5174	0·59	241
Magenta	0·41	+0·59	+0·59	+0·2909	+0·5174	0·59	61
Red	0·3	—0·3	+0·7	—0·1479	+0·6139	0·63	103
Blue	0·11	+0·89	—0·11	+0·4388	—0·0965	0·44	347
White	1·0	0	0	0	0	0	–
Black	0	0	0	0	0	0	–

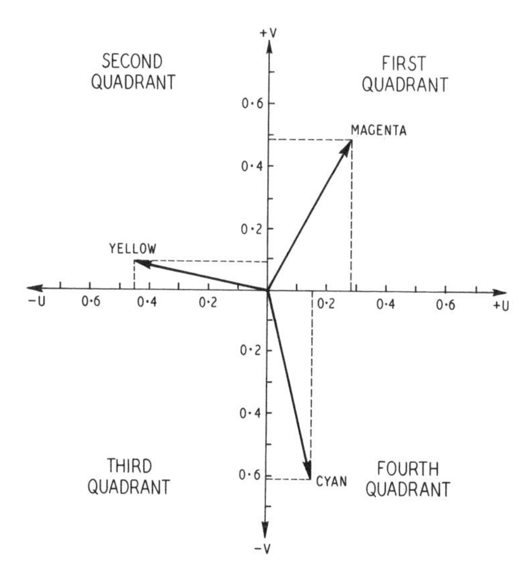

Figure 5.1. Phasors corresponding to the complementary hues of 100 per cent saturation and amplitude

It is now possible to appreciate how the chroma phasor effectively operates over 360 degrees (in the four quadrants), and this is brought out in *Figure 5.1*, which shows the angle and amplitude of the phasor for each of the three complementary colours. This diagram could be extended or another one constructed to reveal these phasor parameters appropriate to the three primary colours.

Chromaticity diagram

Almost any colour can be so represented and ultimately displayed within the wavelength limitations of the colour phosphors of the picture tube. The range of colours is not uncommonly illustrated by the *chromaticity diagram*, an impression of which is given in *Figure 5.2*. Notice the sea of changing hues round the periphery of the horseshoe shape.

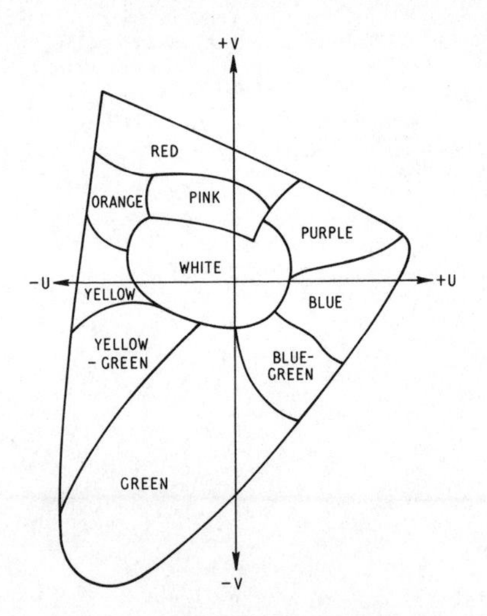

Figure 5.2. Vector axes superimposed on the chromaticity diagram (see text)

This is where maximum saturation exists, the saturation decreasing towards the centre area, where there is no colour. The locus of the horseshoe shape is calibrated in wavelength and a colour is generally described in terms of x and y coordinates (not shown in *Figure 5.2*).

A vector diagram similar to that in *Figure 5.1* is sometimes superimposed onto a chromaticity diagram, suitably modified to take into account the characteristics of the display tube colour phosphors, as shown in *Figure 5.2*. Notice that with the vector axes as shown the complementary hues indicated by the phasor positions in *Figure 5.1* correspond to those same hues on the chromaticity diagram. Another noteworthy point is that the complementary colour of any primary colour lies where a line passing through the achromatic centre (at the intersection of the V and U axes) cuts the locus at two points. Thus we see that the primary of yellow is blue, that of magenta is green and that of cyan is red. This means, therefore, that the magenta phasor (*Figure 5.1*) extended into the third quadrant would correspond to the green angle, that the yellow phasor extended into the fourth quadrant would correspond to the blue angle and the cyan phasor extended into the second quadrant would correspond to the red angle.

Colour clock

To facilitate an understanding of the PAL action the vector diagram is often related to a 'colour clock', as shown in *Figure 5.3*. This carries the primary and complementary hues as the clock 'figures', with the phasor representing the 'hand'. The phasor is drawn pointing to green, which as a primary hue of the PAL system corresponds to 241 degrees, as we have already noted. The greater the amplitude of the phasor, the greater the saturation of the hue to which it is pointing. Thus as each line of picture is scanned, so can be visualised the phasor changing both in amplitude and angle to correspond to the saturation and the hue of the picture elements of each line. When there is no colour the phasor shrinks to zero and the picture is then under the complete control of the luminance part of the signal only.

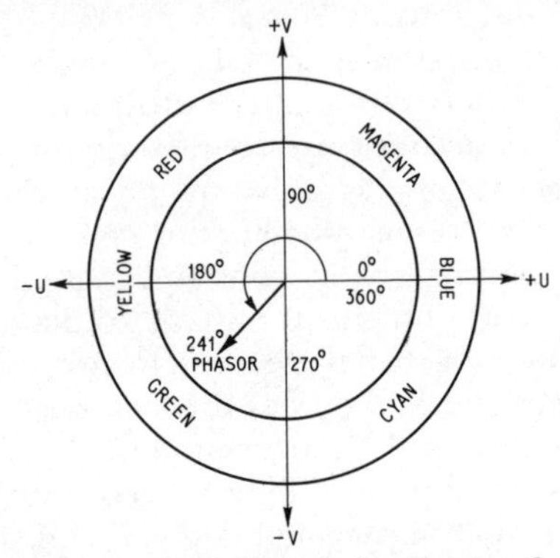

Figure 5.3. 'Colour clock', showing the phasor at 241 degrees corresponding to green

We can now fully appreciate the problem of system phase shift. If something happens somewhere in the system to alter the angle of the phasor then a hue different from that being televised will obviously be displayed. It is not quite so bad if something happens to alter the amplitude of the phasor, for this would not affect the hue but merely the saturation, and a relatively high change in saturation can be tolerated compared with hue. In any case, it is not very difficult to engineer into the receiver a form of chroma signal automatic gain control (a.g.c.), generally referred to as automatic chroma control (a.c.c.).

NTSC hue control

To adjust and correct the phase and hence the hue, NTSC receivers incorporate a hue control which effectively adjusts the angle of the signal until the phasor lines up with the hue or hues transmitted as a reference. A commonly adopted hue reference is flesh tones, and the viewer adjusts the hue control until these appear the most natural in the display.

In practice such adjustments are often required when changing channel or when there is a camera or link change at the transmitter, also should phase distortion occur due to signal propagation shortcomings and abnormal reception conditions.

The PAL system avoids this critical form of manual control by an artifice which counteracts the phase error on one line by introducing an equal but opposite phase error on the next line. In one scheme, called PAL-S, S standing for simple, the eyes of the viewers subjectively average the errors and so the correct hue is automatically discerned in the display. In a more sophisticated scheme, called PAL-D, D standing for delay line, the errors are cancelled electronically before the chroma signals are demodulated.

Both schemes are achieved by the reversal in phase of the V chroma signal for the duration of alternate scanning lines. What happens is that on one line of a field the phase of the V chroma subcarrier is normal, while on the next line of the same field the phase reverses. These can be regarded as 'normal' and 'reversed' lines. While this is happening the phase of the U chroma subcarrier remains normal. *This is not phase reversed on alternate lines.*

The diagrams in *Figure 5.4* show how this corrects the effect of phase distortion. Diagram (a) shows a 'normal' line and phase distortion on a green element causing the phasor angle to lag 45 degrees from the correct 241 degrees. The full-line phasor represents the correct phase and the broken-line phasor (in all drawings) the error phase. The green element on this line, therefore, is displayed towards yellow.

Now, on the next line of the field we have the 'reversed' line, which is shown in diagram (b). Notice here that the effect of the phase reversal is to invert the diagram and to reverse the direction of the phasor, so that the error now leads the correct 241 degrees by 45 degrees. The green element on this 'reversed' line, therefore, is displayed towards cyan.

Diagram (c) clearly shows how the average phase of the two errors (196 degrees on the 'normal' line and 286 degrees on the 'reversed' line) works out to the correct phasor angle for green, which as we have seen is 241 degrees.

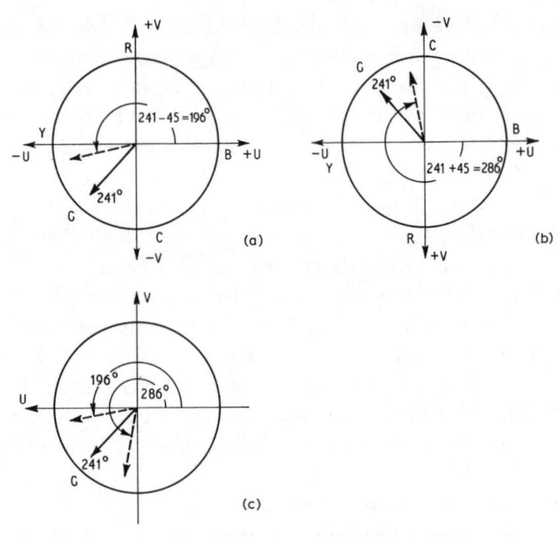

Figure 5.4. Diagrams revealing in simplified form the phase error combating artifice of the PAL system. (a) *phasor due to green element occurring at 196 degrees instead of 241 degrees on a 'normal' line.* (b) *the same phase error on a 'reversed' line. The error is now reversed, and diagram* (c) *shows that the average of the two errors is 241 degrees, corresponding to the correct phasor angle for green*

The green sensation of the two errors is provided by a PAL-S receiver, as shown in *Figure 5.5*. The function is thus subjective. PAL-D receivers arrange for the averaging to be done electronically, and owing to this greater phase errors can be accommodated. One problem of the PAL-S subjective averaging is that horizontal lines appear on the picture when the phase error exceeds a relatively small value. This is

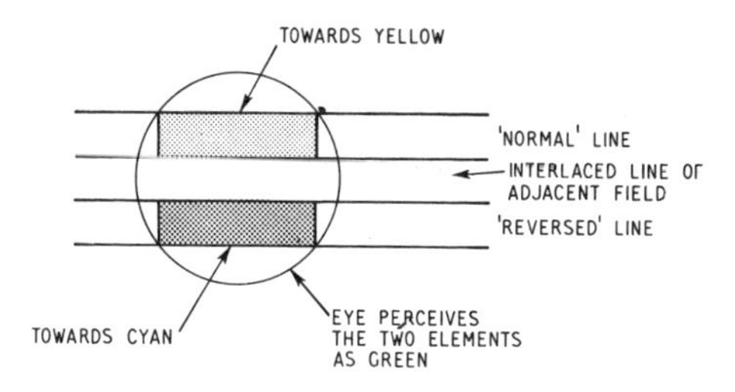

Figure 5.5. PAL-S receivers show the element as towards yellow on the 'normal' line and as towards cyan on the 'reversed' line, the eye then averaging the two colours and perceiving the correct green element

because successive lines of a field do not lie side-by-side, since the lines of one field are interlaced between those of an adjacent field (see Chapter 4). The subjective averaging thus tends to occur over a picture composed of half the number of active lines, and when the phase error rises the eye discerns the line structure carrying the opposing hue errors. The resulting horizontal line interference is known as the *Hanover blind effect*, since it was at Hanover (West Germany) where it was observed during the development of the PAL system at the Telefunken laboratories.

V chroma detector switching and synchronising

Some readers may now have become aware that owing to the V chroma subcarrier phase reversals, the R − Y output from the V chroma detector would also reverse in phase unless counteracting switching was introduced at that particular detector. PAL receivers, therefore, incorporate an electronic circuit to switch the phase, either of the reintroduced subcarrier (called the *reference signal*) or of the V chroma signal applied to the V detector. The actual switching operation is not difficult since the switch can be activated by pulses from the line timebase. More difficult are the circuit and control required to ensure that the detector switching is synchronised to the 'normal' and 'reversed' lines, for clearly if the detector was switched to work from a 'normal' line when the

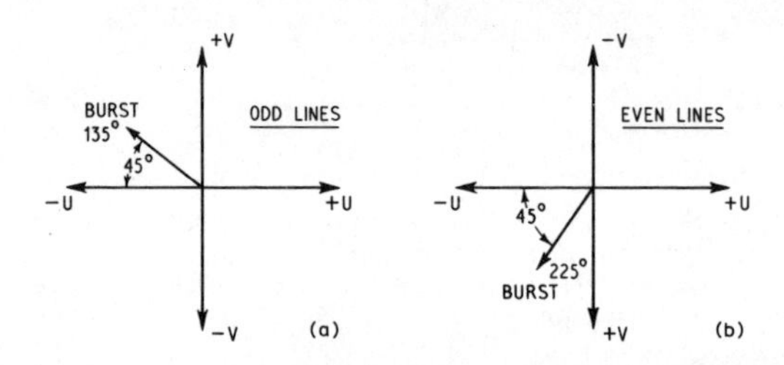

Figure 5.6. Showing how the bursts swing ±45 degrees relative to the −U axis on successive lines. That is, from 135 degrees on odd lines to 225 degrees on even lines. This means that the average phase of the bursts is 180 degrees, coincident with the phase of the −U chroma axis. Notice how the burst swings are geared to the V chroma phase alternations

input was a 'reversed' line the displayed colours would be significantly in error!

This synchronising is handled by the colour bursts of the PAL signal. It will be recalled that the fundamental purpose of the bursts of the NTSC signal is to frequency- and phase-lock the reference signal at the receiver (other functions of the bursts include colour killer switching and automatic chroma control). The PAL bursts, however, are caused to swing in phase 45 degrees either side of the −U chroma axis in synchronism with the phase switching of the V chroma subcarrier. This is revealed in *Figure 5.6*, where the burst phase on a 'normal' line is indicated at (a) and that on a 'reversed' line at (b). Clearly, the *average* phase of the bursts is coincident with the −U chroma axis (180 degrees), and it is to this which the reference signal at the receiver locks, which is the requirement for correct colour reproduction on a PAL receiver.

The NTSC bursts do not swing in phase like this; the phase is fixed to that of the B−Y axis, with the I axis 57 degrees in retard and the Q axis in quadrature with the I axis.

Figure 5.6 shows the 'normal' lines corresponding to odd-numbered lines and the 'reversed' lines to even-numbered lines, allowing us to refer the V chroma subcarrier and burst phases to odd and even lines. Thus from *first principles* on, say, line 7 of a field we get the conditions at (a), on line 8 the conditions at (b), on line 9 the conditions at (a), and so on.

More about Hanover bars

Remembering that the full number of lines of a complete picture is made up of the interlaced lines of *two* fields, it will now become more apparent why Hanover bars occur when the phase error is significant on PAL-S receivers. The

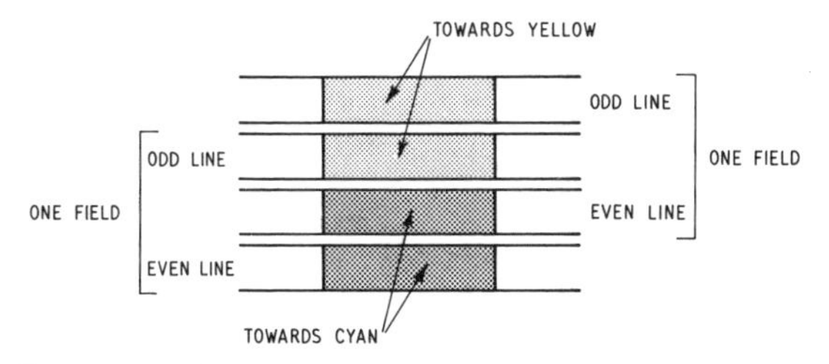

Figure 5.7. Owing to interlaced scanning, pairs of adjacent lines carry the real and opposing hue errors, and when the error is large this results in the display of Hanover bars (also see Figure 5.5)

subjective effect illustrated in *Figure 5.5* applies to one isolated field only. When the lines of two fields are properly interlaced the effect shown in *Figure 5.7* results. Here two adjacent lines of a complete picture (frame) carry the hue error, while the next two adjacent lines carry the opposing hue error. The eyes are less able to integrate the two pairs of lines in terms of the true colour, hence the bars.

V chroma phase identification

Because one switching cycle occupies two lines, the switching occurs at half line-frequency, which on the 625-line system is about 7.8 kHz. Signal at this frequency is generated in the receiver due to the swinging bursts and it is this which identifies the odd and even lines at the V chroma detector for chroma phase or reference signal phase switching.

The signal emanates from the 'phase detector' which is a circuit that provides the reference signal with a phase lock from the bursts proper. More about this is said in Chapter 9. The 7.8 kHz signal is sometimes referred to as 'ident' (short for identification) signal so far as its application for V chroma line identification is concerned. The signal might also be used for colour killer switching and, in very few cases, for automatic chroma control. In such applications it might be known as 'ripple' signal. The ripple signal in some receivers also activates a chroma rejector circuit in the luminance channel, meaning that the rejector is out of circuit on monochrome transmissions and thus fails to impair the picture definition, which it might do due to bandpass restriction if it remained in circuit on monochrome transmissions. Automatic chroma control is mostly provided by a bias which is derived from burst rectification.

The block diagram in *Figure 5.8* illustrates the basic PAL transmission system. Because the luminance signal operates within a wider band than the chroma signals, a delay network is introduced in the luminance feed to the adder. This ensures that the luminance signal arrives at the transmitter exactly at the same time as the chroma signals. Without this the luminance signal would arrive early, because signal is less delayed as the bandwidth of a circuit is increased. A reciprocal delay is incorporated in the receiver, but it has nothing to do with the chroma delay line, the function of which is explained in Chapter 9.

The diagram also shows the 90-degree (quadrature) shift in the subcarrier circuit between the V and U chroma modulators. Remember that weighting factors are applied in

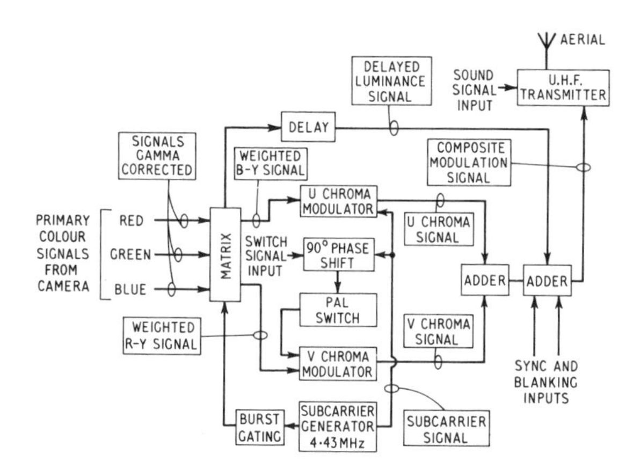

Figure 5.8. Block diagram of basic PAL transmission system

the B−Y and R−Y channels to the U and V chroma modulators. The PAL switch, providing the alternate line phase reversals of the V chroma signal, is also shown in the diagram.

The oscillogram in *Figure 5.9* shows one line of composite signal including the burst but minus the line sync pulse which occurs on the standard colour bar transmission.

We have seen, therefore, that in essence the PAL signal differs from the NTSC signal in that alternate line phase reversals of the R−Y subcarrier and synchronised swings of the phase of the colour bursts are included as a method of cancelling signal phase error and hence hue error in the

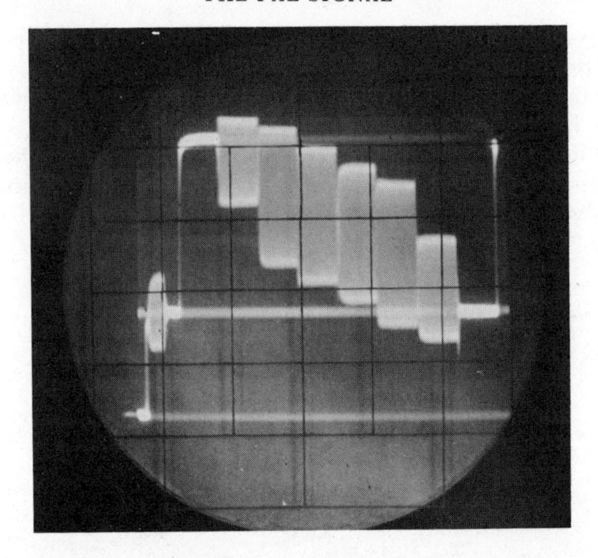

Figure 5.9. Oscillogram showing the composite signal (but excluding line sync pulse) on one line of a colour bar transmission

display. There are other differences, including the V and U axes against the NTSC I and Q axes and the reduced bandwidth of the chroma signal on the Q axis (about 0.5 MHz) relative to that on the I axis (about 1 MHz).

Gamma correction

To conclude this chapter, mention should be made of the *gamma correction*, indicated in *Figure 5.8*, of the primary colour signals fed to the matrix (an electronic adder/subtractor system). Such correction is applied to the camera signals to correct for the non-linearity essentially of the picture tube. Without correction the picture highlights would be stretched and the low-key levels would be compressed, resulting in a display of non-linear contrast ratio. This is because a television picture tube fails to produce light in direct proportion to the signal voltages applied to it, which means that the light output of the picture tube is not proportional to the light input at the camera. Gamma correction thus solves this problem.

6

COLOUR TRANSMISSION

The obvious way to send electrical information over long distances would seem to be by using cables, which is in many cases possible. One of the drawbacks to this method, however, is the distortion produced by the cables when they are required to handle a wide band of frequencies. Amongst the electrical characteristics of a cable are resistance, inductance and capacitance, and since the last two of these are frequency conscious, there will be differences in the characteristics at different frequencies.

Communications engineers solve the problems associated with this by inserting 'correcting' units, often called 'equalisers', along the cables at calculated distances. To use a cable network for the distribution of colour television signals the simplest plan would be to arrange for three separate circuits, thus eliminating the need for encoding and decoding. This is a solution only if a few receivers are to be served by the system, and the technique has been adopted in closed circuit television and certain modes of 'relay' television.

However, when a large number of receivers has to be catered for it is technically less of a problem to take the three primary colour signals, encode them and add them to the luminance signal for transmission in the manner so far described. The final composite signal may then be distributed through a single cable system or transmitted in the usual manner. Some engineers consider that cable distribution is the best method to use, particularly for local distribution of the signals since it avoids each household having to have its own aerial system. There can thus be fewer problems of interference. However, for signal propagation over tens of miles the best method is still from a television transmitter, with each receiving point collecting the signal from a smaller aerial of characteristics to suit the prevailing signal and interference conditions.

Space transmitters

It is highly likely that within the next decade or two world-wide television signal propagation will be by way of space satellites, with each household either having its own small 'dish' aerial to collect the signals from the space station or being wired to a local cable distribution system, with the input of the system collecting the space-borne signals from a more elaborate aerial than could be accommodated at each household. Already quite a bit of background work has been done on this mode of signal propagation and, of course, it is already being used for international colour television links and space projects. The signals will be of far greater carrier frequencies than employed so far for domestic television, but the aerial used at each household might be equipped with a small transistorised converter to translate the super high-frequency received signals to lower frequencies more readily acceptable by domestic television receivers. There is still a lot of work yet to be done to bring this about, and for the time being our colour signals will continue to come to us via the ultra-high-frequency channels of Bands IV and V and chimney-mounted aerials, except in those areas employing cable distribution systems (see *Beginner's Guide to Television*).

In order to send information over relatively long distances it is necessary to modulate a carrier wave, and the propagation distance and 'quality' of the received signals are somewhat governed by the choice of carrier frequency. Even some cable distribution systems employ modulated carriers for the television information, and in this respect the cables can be regarded as transmission lines which to some extent act as does the ether (space) to radio signals.

As we have seen, colour television contains information extending over a band of frequencies up to 8 MHz, including the accompanying sound information, when one of the sidebands is partially suppressed (see *Figure 4.8*). If this information were amplified to a very great power and then fed direct to the transmitting aerial the information contained in the frequencies at the high-frequency end of the spectrum might reach the receiver, but it is certain that the information at the lower frequencies (virtually down to 0 Hz or d.c.) would never arrive. Because all the frequencies must be present to secure a complete and undistorted picture (and sound) this state of affairs would not be satisfactory.

Need for modulation

The problems are neatly solved, of course, by modulating a very much higher frequency carrier signal with the information to be transmitted, subsequently extracting it at the receiving end by a technique known as 'detection' or 'demodulation'.

The higher the carrier frequency the smaller the 'gap' occupied by the carrier and its information-carrying sidebands obtained from the modulation. This is true whatever the information, whether it be sound, vision, computer code, data, etc.

Bands and channels

Colour television in the U.K. is transmitted in the u.h.f. (ultra-h-frequency) band. This is divided into two bands, called Bands IV and V and each band is divided into a number of 8-MHz channels. In fact, Band IV has Channels 21 to 34, from 471.25 MHz to 581.25 MHz, while Band V has Channels 39 to 68, from 615.25 MHz to 853.25 MHz. The sound carrier in each channel is separated from the vision carrier by 6 MHz. Each local area in the U.K. has been allocated four u.h.f. channels, three of which are in current use for the programmes of BBC1, BBC2 and ITV1. The future of the fourth channel of each group has not been determined at the time of writing, though there are some indications that it might be adopted for a second ITV channel, giving ITV2; but this is by no means definite.

Each local group of channels works within a given u.h.f. spectrum, depending upon the circumstances, and many of the groups have a spectrum extending over 88 MHz. Some groups have a wider spectrum, and each group is known by a letter and colour, mainly to facilitate aerial grouping, since aerials are now designed to respond equally to the three channels of a group. Thus a viewer of the 625-line pictures in the u.h.f. bands needs only one aerial to receive the three programmes. The transmitters of the three programmes, of course, are co-sited to make this possible. Co-siting and wideband aerials were not feasible in the earlier v.h.f. bands (Bands I and III for television, with Band II for frequency modulation sound), so two or more aerials were required to obtain the various programmes.

Table 6.1 gives the groups in terms of channel numbers.

Table 6.1 AERIAL GROUPS

Group	*Colour code*	*Channels*
A	Red	21–34
B	Yellow	39–51
C	Green	50–66
D	Blue	49–68
E	Brown	39–68

Viewers should thus make sure that the u.h.f. aerial used corresponds to the local channel group. If the incorrect aerial is used the signals from the various channels will be out of balance and severe picture distortion, with reflections and 'ghosting' effect, may result.

Types of modulation

The four types of modulation used in communications to attach information to a carrier wave are (a) amplitude modulation, (b) frequency modulation, (c) phase modulation which is a species of (b), and (d) pulse-code modulation. In ordinary sound broadcasting (a) or (b) is mostly adopted. In the earlier 405-line monochrome television system (a) was used for both sound and vision transmission, but in the current 625-line system (a) is used for the vision modulation and for the chroma modulation in conjunction with the suppressed subcarrier technique, as has already been discussed, and (b) for the sound signal.

Pulse-code modulation

Although (d) was originally devised in the mid-1930s by Alec Reeves, it is only of relatively recent date that its use has been considered for sound transmission in terms of the conveyance of audio information over radio (and other) links of specific bandwidth without impairment to the quality of the signal or to its signal-to-noise ratio. It seems likely at the time of writing that this kind of modulation will be adopted for country-wide sound broadcasting links between programme sources and transmitters instead of the other kinds of modulation, or direct cabling, previously employed. This will allow two audio channels to be accommodated per programme with less strain on the available 'radio space' which, it is hoped by many enthusiasts, will encourage the transmission of more stereo programmes via the normal medium of frequency modulation sound broadcasting with stereo multiplex.

So far as television is concerned, pulse-code modulation is being seriously considered for adding the sound information to the actual vision signals as a means of reducing the costs and enhancing the capacity of the coaxial cable and/or microwave links between studio centres and transmitting stations. Indeed, the scheme, called 'sound-in-vision', was first tried by the BBC Research Department on the 625-line system during the latter part of 1968 with encouraging results. The trial was over a 760-mile loop extending from London to Scotland and back. The sound signal was combined with the vision signal at London and the composite signal was sent over the vision link to the Kirk o' Shotts station, where the sound and vision signals were separated and directed to the sound and vision transmitters at Black Hill for transmission in the usual manner.

Pulse-code modulation is a highly complicated subject, which cannot for obvious reasons be explored in a beginner's book of this kind. However, the function is such that the information is coded into digital terms by using a digit code of specific nature. It is then sent over the link in this form and at the receiving and translated back to the original form. It thus constitutes analogue/digital coding at the sending end and digital/analogue decoding at the receiving end.

The BBC's tests mentioned employed the 4.7 microsecond line-synchronising intervals of the vision signal to accommodate the pulse-code modulation for a period of 3.8 microseconds. A great advantage of this kind of modulation is that interference on the system, and this includes general 'noise', fails to affect the digital/analogue translation in the same way, which means that distortions and noise on the link have no effect at all on the decoded sound signal. Perhaps when the next edition of this book is written there will be something more definite about the exploitation of this kind of modulation for television applications.

Amplitude modulation

We saw in Chapter 4 that in amplitude modulation the instantaneous value of the high-frequency carrier wave is varied by the lower frequency information signal. The signals involved and the resultant modulated carrier wave are illustrated in *Figure 4.4*.

Frequency modulation

As already intimated, this type of modulation has come to be used increasingly over the last decade for high-quality (hi-fi) sound transmissions in Band II, and it is also geared into the multiplex system of stereo transmission and reception (see *Beginner's Guide to Radio*). It is also used for the transmission of the television sound signals on the 625-line

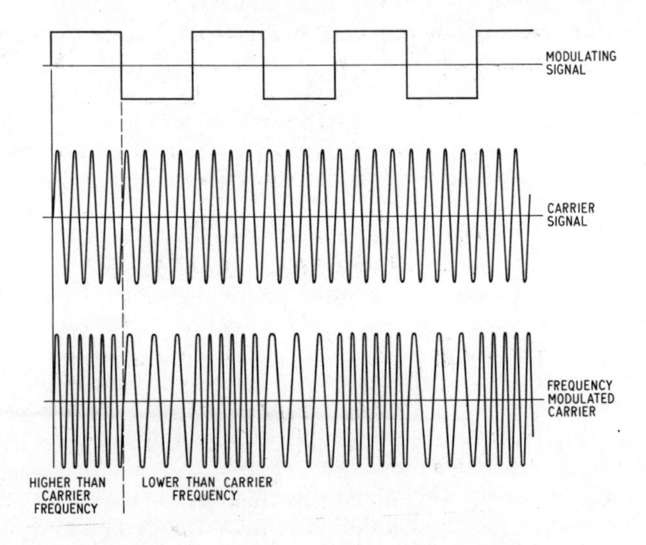

Figure 6.1. Illustrating frequency modulation

system (amplitude modulation, though, was used in the 405-line system). In this type of modulation the carrier amplitude is kept constant and the frequency of the wave is varied at a rate depending on the frequency of the modulation signal and by an amount (deviation) depending on the strength of the modulation signal (see *Figure 6.1*). Thus, the 'louder' the sound, the greater the deviation, while the higher the frequency of the sound, the greater the rate at which the carrier frequency is varied either side of its nominal frequency. Maximum deviation of the BBC's sound system is ±75 kHz (about ±50 kHz for television sound). This corresponds to the maximum modulation depth; thus commonly referred to as 100 per cent modulation. 100 per cent amplitude modulation is when the troughs of the modulation signal forming the envelope of the modulated waveform fall to the zero datum line.

Phase modulation

Here the phase of the carrier wave is altered according to the modulating signal. It has much in common with frequency modulation, and it is sometimes regarded as the 'nature' of the modulation of the chroma signal resulting from quadrature amplitude modulation by the R−Y and B−Y signals and subsequent suppression of the subcarrier. It will be recalled that the phasor angle (e.g., phase) of the signal is governed by the colouring information.

Sidebands again

It has been shown (Chapter 4) that whatever type of modulation is used, sidebands are produced. Treated mathematically, the modulated signal can be shown to become a signal of single frequency (the carrier wave) on either side of which are spread a number of signals whose frequency and amplitude depend on the excursions of the modulating signal.

This results in a 'spread' of signals across a frequency band, as shown in *Figure 6.2*.

These are the sidebands, and the maximum 'spread' is determined by the highest frequency to be transmitted as modulation. Thus a high-frequency modulating signal

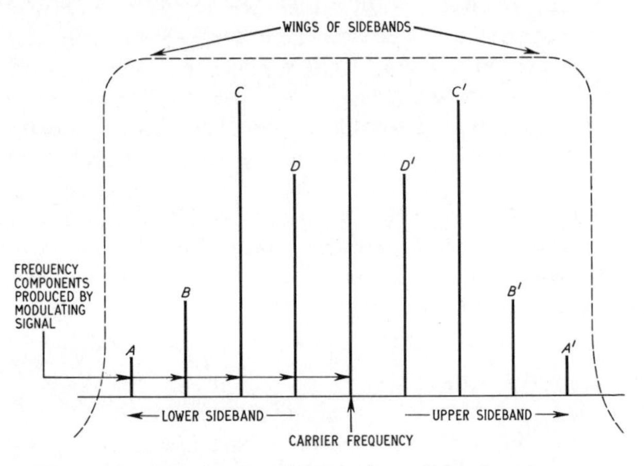

Figure 6.2. Showing the sidebands of a modulated carrier wave

will produce sidebands which are located at the extreme wings of the bandwidth. These high-frequency modulating signals correspond to the fine detail in the picture. Consider, for example, a mole on an actor's face; a black mole on a 'white' face which is already small compared with the entire picture area. The change from white to black and to white again can be regarded as consisting very roughly of a half cycle of sine wave signal. Now, bearing in mind that the number of such waves that could be accommodated in a single line of picture would be quite large, that a complete picture is composed of more than 600 active lines and that 25 complete pictures are produced each second (e.g., 50 fields of half the number of lines per second, with interlacing giving the full number of lines per complete picture), it does not require very much effort to calculate the frequency that must be handled to yield the fine detail of our actor's mole! It could be around the 4 MHz mark. More information on this topic (not of the mole!) is contained in *Beginner's Guide to Television*.

Conversely, if the same calculation is performed for the large area formed by, say, a black cat, it will be found that the frequency is very much lower. The lower frequency components contain the 'main body' of the information, and so are also very important. Later in this chapter the need to retain the low frequencies by using vestigial sideband transmission is explained. To summarise, therefore, the fine detail which gives a picture 'sharpness' and clarity is produced by the high modulating frequencies which develop on the wings of the sidebands in *Figure 6.2*. The vertical lines in this drawing each represent a single frequency signal of a frequency and those A′ above are the result of the carrier being modulated with high-frequency signal, say, 4 MHz, whereas D and D′ correspond to a much lower frequency, with C/C′ and B/B′ corresponding to intermediate frequencies.

Bandwidth

This band of frequencies has to pass through the various types of equipment, including amplifiers, aerial systems, detectors and so forth in its passage from the studio or programme source to the viewer's home. The bandwidth of such devices will determine just how much of the detail of the original signal gets through. If the bandwidth is too small it will eliminate some of the 'detail' information and also, perhaps, distort the signal in other ways. If on the other hand the bandwidth is too great it might allow undue 'noise' and other unwanted signals to enter the system and thus detract from the inherent quality of the vision (and sound). In any case the signal must not spread too far over too wide a bandwidth because there are other signals from other transmitters which need to occupy channels near to it. Mutual interference would thus be likely to occur if the bandwidth is allowed to overlap into adjacent channels. A very neat 'model' illustrating these points is given in *Figure 6.3*.

Carrier suppression

Because the information effectively travels in the sidebands it is possible to suppress the carrier signal altogether and transmit only the sidebands. This scheme, it will be recalled, is adopted with respect to the chroma signal.

It has also been mentioned (Chapter 5) that in order for the sidebands to release their information it is necessary for the suppressed carrier to be regenerated at the receiver. The reference signal constitutes the missing subcarrier—*sub* because it is a secondary carrier below that of the main carrier wave—and the frequency of this is arranged accurately to match that of the original subcarrier. Since some of the information in the chroma signal is in the form of 'phase variations' (the phasor) it is highly essential for the phase of the regenerated subcarrier also to be very accurately maintained and to match that of the subcarrier. Frequency and phase synchronising is performed by the colour bursts, which are as we have seen bursts of subcarrier generated on each of one of the black level intervals to the line sync pulses.

Interleaving

A fascinating thing about colour television is that on paper it would appear almost impossible to work satisfactorily without enormous expense and difficulty, yet in practice it performs virtually without flaw and with relatively few problems. At first sight it might seem that the bandwidth required to transmit a compatible signal would be impossibly

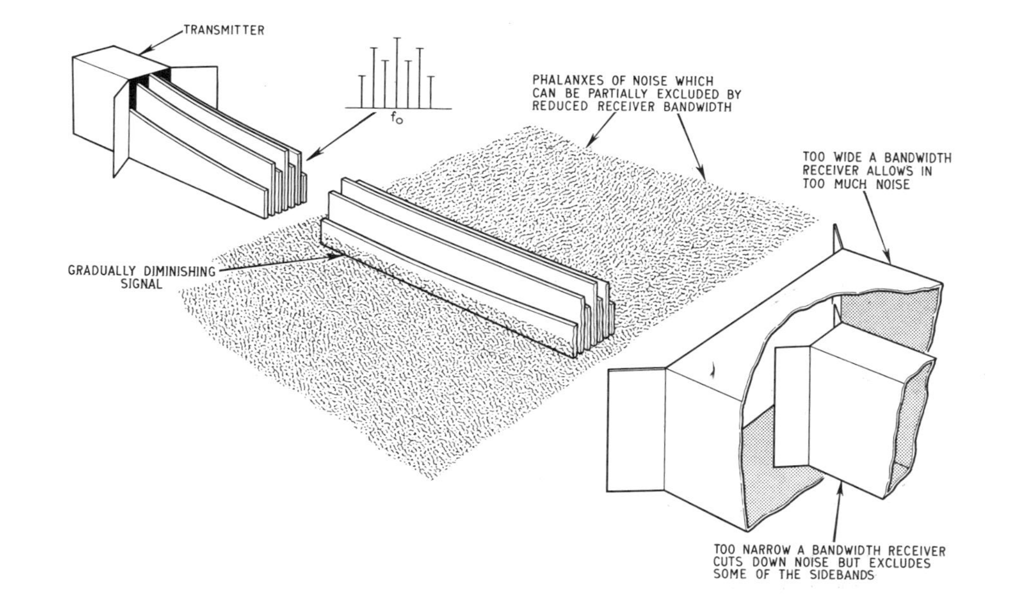

Figure 6.3. Model showing the relationship between the sidebands and the bandwidth

large, with the luminance sidebands filling one space and the chroma sidebands extra space alongside. However, we should now realise that this is not true because the system is cleverly engineered so that the sidebands of the chroma signal interleave with those of the luminance signal.

Two American engineers, Mertz and Grey, were amongst the first to observe this interleaving effect. Previously it was assumed that the whole sideband space on either side of the carrier frequency was completely filled with essential information, but the two Americans were able to demonstrate that the majority of the important information occupies only discrete bands, leaving gaps which are hardly ever filled with sideband energy. The position of the gaps is related to the number of picture lines and hence the line timebase frequency. A colour television system thus relates the subcarrier frequency to the line timebase frequency so that optimum interleaving occurs. Some idea of interleaving was given in *Figure 4.9*, but a more formal illustration of the effect is given in *Figure 6.4*, which also shows the vestigial

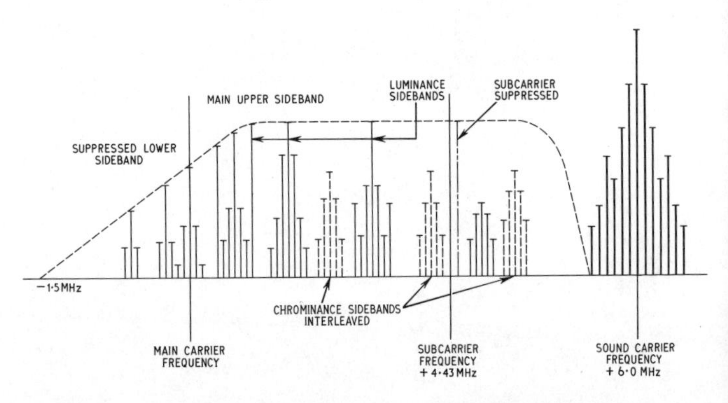

Figure 6.4. Detailed diagram showing how the sidebands of the chroma signal are interleaved with those of the luminance signal. The diagram also shows the sidebands of the sound signal and the suppressed lower sideband

sideband characteristic of the signal, referred to again at the end of this chapter.

The significance of interleaving is that all the information, luminance and chroma, can be sent within a channel no wider than that needed hitherto for monochrome only.

To summarise, then, the interleaving is achieved by correctly relating the subcarrier frequency to the line scanning frequency, and this is done at the studio centre where a common 'master' signal source provides the timing for both the subcarrier signal and the line sync pulses. In Chapter 4 the actual frequency of the subcarrier in the British PAL system was given as 4.433 618 75 MHz, and this is related to the line frequency thus:

$$\text{Line frequency} = \frac{f_{sc} - \frac{1}{2}f_f}{284 - \frac{1}{4}}$$

where f_{sc} is the subcarrier frequency and f_f is the field timebase frequency which, in the British system, is 50 Hz.

Transmitter links

The composite signal from a studio centre is sent over a cable or radio link to the transmitting station, which may be many miles away in some remote and sometimes almost inaccessible place, not uncommonly at high altitude on the top of a hill or mountain. When cable links are employed careful equalisation is necessary to ensure that the full spectrum of the signals is accurately maintained. Vision (or video) signals have a much greater bandwidth as we have seen so the link system must be capable of handling these without low- or high-frequency suppression or distortion. The sound signal presents less of a problem since its bandwidth extends to a maximum of only 15 kHz, compared with 5 or 6 MHz of the video signal.

A radio link consists of a microwave transmitter and receiver over which both the sound and vision signals may be sent, though sometimes the sound may go via specially prepared (equalised) Post Office lines. The aerials at each end of the link are of the 'dish' type, not unlike those used in some radar installations. The transmitting aerial is very accurately beamed onto the receiving aerial, which is sometimes seen located about half-way up the main aerial mast at the transmitting site.

The studio centres are commonly placed in the middle of large towns or built-up areas, and special Post Office lines are laid on to direct the sound and vision signals from the studio equipment to the cable or radio link(s). The studio centres, of course, are also wired for inter-control, etc. The input signal may be 'live' from a camera or derived from a film or video tape, the latter 'storage' medium being frequently used nowadays. The signals are directed through a control centre to all the 'networked' transmitters via the link system.

The use of pulse-code modulation in the links was mentioned at the beginning of this chapter.

Transmitter range

At u.h.f. the distance over which signals can be transmitted is limited under normal reception conditions to about 30 or 40 miles, depending on the height of the transmitting and receiving aerials. This fact is due to the nature of these ultra-high-frequency signals—they travel straight outwards and do not bend so easily round the Earth's surface and obstructions as do the lower frequency signals.

If one stands on the top of an aerial mast, then as far as the eye can see on a very clear day will be just under the limit of consistent reception. Reception is possible slightly beyond the horizon distance because of the height of the receiving aerials and because of *slight* bending of the signal round the curved Earth. This is why transmitting stations are located on chosen sites high above sea level clear of interference.

Relay stations

To 'illuminate' those areas out of range of the main transmitters or shaded by local hills or mountains, 'relay' transmitters are used. These are lower power than the main ones and often automatic in operation. Channel group conversion is built into them to avoid the possibility of the main and relayed signals on the same frequency causing interference.

Super high-frequency radio links may be employed to bring in the signals for re-radiation or the nearest main station may be received on high-gain and special aerials located high up the mast, 'cleaned' and amplified and then changed in channel grouping before re-radiation from the relay transmitter.

Vision transmitter

Each station usually has a 'standby' transmitter which comes into operation should the main one fail. Until the very end of the chain the sound and vision signals are kept completely separate, and it is not normally until a point just before the aerial that they are combined. With pulse-code modulation, however, the sound signals are deliberately integrated with the vision signals, but this is not quite the same thing.

Some transmitting stations include basic studio equipment for producing test signals or in case failure in the links or studio centre necessitates the station originating its own programme material which, in most cases, is rarely more than test cards and slides, the latter to inform viewers what is happening or apologising for the interruptions of the normal programme.

The video signals from the studio centres are received, monitored and then 'clamped'. This latter procedure is very important in television. The various components of the composite signal have to be maintained in specific proportion. The black level intervals, for example, have to hold steady at all times, as does the amplitude of the sync pulses and colour bursts. Variations in proportions and levels can occur in the signal after leaving the studio centres, and so it is arranged for the transmitter to clamp the signal electrically to a stable and fixed level before it is applied to modulate the carrier wave which, of course, is the fundamental yield of the transmitter itself.

The carrier wave frequency has to be very accurately maintained as well, and quite a lot of the transmitter equipment is devoted to the birth of this signal from a very stable crystal oscillator. The crystal produces a signal of much lower frequency than required for the carrier wave and multiplying circuits are used to step up the frequency to the required channel. The crystal is housed in an oven which keeps its temperature constant, which is necessary for a constant-frequency output.

The u.h.f. signal is then passed through a number of stages of amplification which raises its power to many thousands of watts, the signal then being coupled to the aerial system, whose gain increases the effective radiated power. This results in remarkably high power, the London Crystal Palace u.h.f. transmitter, for example, having an effective radiated power of 800 000 watts (800 kW).

The composite signal from the studio centre is also amplified and then fed at high level to the radio-frequency transmitter to amplitude modulate the carrier wave in the manner already described.

Sound transmitter

The sound signal is also brought into the transmitting station and monitored before being fed to the preamplifiers which supply the modulation input to the sound transmitter.

Originally in the black-and-white 405-line system the

sound was carried by amplitude modulation, the same as the vision. In the present colour system, however, frequency modulation is used, which has several attributes over amplitude modulation, for sound anyway. The dynamic range of the audio information (e.g., the range between the softest and the loudest sounds) can be more readily retained, and because most of the interference experienced on sound reception (and vision come to that) results from amplitude variations of the modulated waveform, this can be eliminated by amplitude limiting (e.g., shaving off the top and bottom of the waveform) at the receiver when the sound is carried by frequency modulation. Moreover, high-quality sound reproduction is possible from a properly engineered f.m. sound section, and frequency compression is not needed to reduce the effects of interference and noise, as it might be when amplitude modulation is used for the sound signal.

Other desirable features stem from f.m. sound in the system as a whole. For example, it makes possible at a receiver a common i.f. (intermediate-frequency) channel for the sound and vision signals, owing to the lack of interaction between the two types of modulation. It also allows the system of intercarrier sound to be used, where a 6-MHz f.m. sound signal is produced from the beat between the sound and vision carriers (because these are separated from each other by 6 MHz). Since the accuracy of this 6-MHz signal is reflected by the crystal-controlled accuracy of the sound and vision carriers produced by the transmitters, it follows that the intercarrier sound system significantly reduces the effect of frequency drift at the receiver on the sound signal, compared with the obtaining of this direct from the sound carrier.

Signal-to-noise ratio

Radio-man has over the years endeavoured to secure the highest possible ratio between the wanted signal and the unwanted noise. The signal for transmission may start out with a high signal-to-noise (S/N) ratio, say, at least 10 : 1, which means that the wanted signal is ten times stronger than the unwanted noise. However, during the process of amplification, etc., the S/N ratio diminishes. This is because amplification actually adds to the original noise signal. Indeed, all passive and active components in the system add to the noise, so the greater the route taken by the signal through the various items of equipment, the greater will be the noise added to it when it emerges.

There are various ways that the engineer can ensure that the noise is kept to the least possible. One is by making the bandwidth of the circuits as narrow as possible without excluding any of the components of the wanted signal. It is not possible of course to narrow the vision bandwidth very much, but there is certainly no need for the sound bandwidth to be as great as that of the vision, for the sound carries far less information.

For this reason the power required to transmit the sound signal need not be so great as the vision. Think of it this way: if significant noise is present outside the transmitter, waiting for any signals to leave, then the signal with the widest band is going to pick up more noise than the one with the narrow band. 'Open the window wider and more dirt comes in', as the audio chaps would say!

The vision transmitter is thus made more powerful than that for the sound signal. A ratio of 5 : 1 (International Telecommunications, Geneva recommendation) is commonly adopted.

Vision/sound combining

A common aerial is employed to radiate the sound and vision signals, which means that the outputs from the transmitters have to be combined. The frequency of the sound carrier is higher than that of the vision carrier in the 625-line system (it used to be the other way round with the 405-line system) and the separation is 6 MHz, as just noted.

The actual combining of the two signals is not all that problematical. What does cause trouble, though, are the mutual effects of the aerial and the two transmitters on each other. They could interfere with each other's working unless very carefully matched. Lack of matching results in the return of a large amount of energy, and the return of, say, sound energy to the vision transmitter could cause embarrassment. The problem is solved by a special combining unit, as shown in *Figure 6.5*.

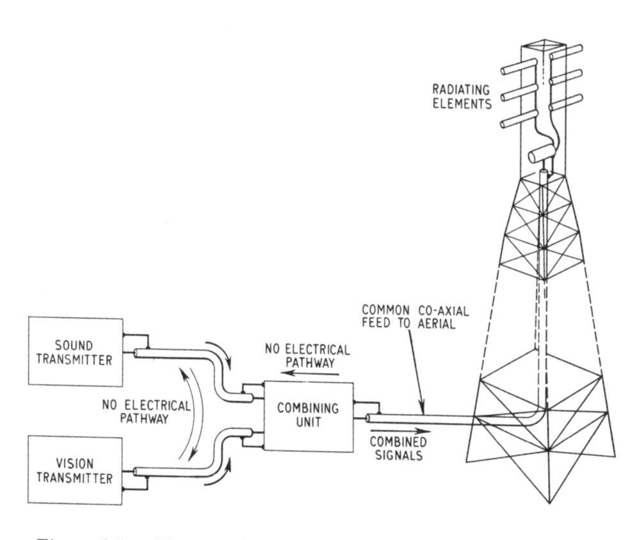

Figure 6.5. Showing the action of the sound/vision combining unit at the aerial

The term 'matching' in electronics implies that two or more circuits have been very accurately tailored so that maximum coupling, with least signal reflection, occurs between them. This can be shown to occur when the external load impedance is exactly equal to the internal impedance of the power source. In communications it means the absorption by the load of all the energy actually sent out by the high-frequency power source—radio transmitter or amplifier—which in the present case is the aerial system. The matching may be achieved in a variety of ways, but the purpose is the same as in the most elementary case; that is, the external impedance is made to 'look' in the electrical sense exactly the same as the internal impedance of the source and so satisfy the conditions for the transfer of maximum power from the generator to the load. This is just what the combining unit does for the sound and vision transmitters.

Aerial system

The aerial system consists of the actual radiator arranged as a stack of conductive elements at the top of the transmitting mast and the feeder and coupling system which directs the transmitter power to the radiator (called aerial array) at the top. The arrays are usually more complicated that those used for the reception of the signals, though the basic principle is the same—reception in the aerial sense is merely the reciprocal of transmission.

Instead of there being just one active element, as there is in most domestic aerial arrays, there are several of them which, with passive elements, yield a pattern of radiation most suitable for the location. It is better, for example, to have as much of the energy as possible directed at the horizon rather than into space! If the transmitter is by the sea, then the energy is generally required inland and not out to sea. *Figure 6.6* shows at (a) the geometric pattern of a line-of-sight transmission path, while (b) illustrates the fall-off in signal strength with distance from the aerial. In the early 405-line system most transmitters radiated vertically polarised signals (e.g., with a plane of vertical polarisation);

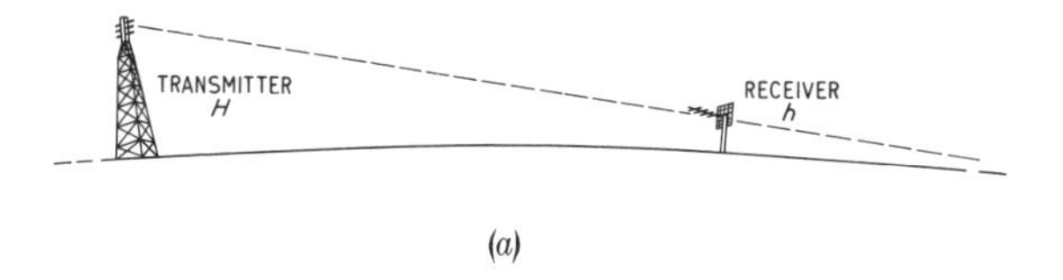

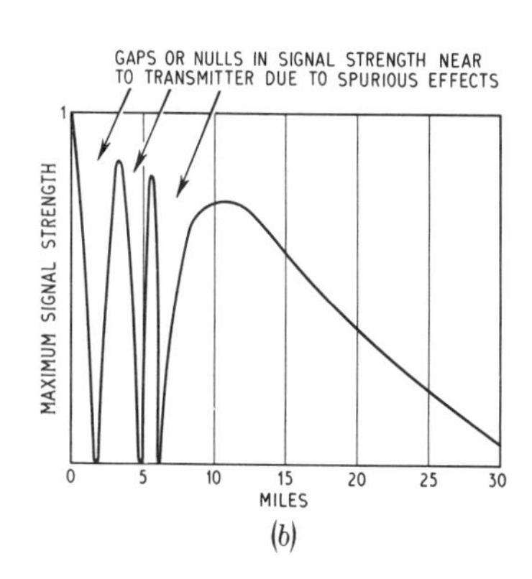

Figure 6.6. (a) *line-lf-sight path of aerial at u.h.f.* (b) *curve showing how the signal field can vary with increasing distance from the aerial. The violent changes at distances close to the aerial are caused by interference effects resulting from signal reflections from the Earth. After this the signal intensity reduces with distance*

that is, with the electric field of the wave in the vertical plane, this calling for the receiving aerials also to be mounted with the elements vertical. Many of the u.h.f. stations on the 625-line system adopt horizontal polarisation. Polarisation is dealt with more fully in Chapter 8, where the domestic aerial is investigated in detail.

Aerial gain and directivity

There is a property about aerials, called *gain*, which has an important bearing on signal strength. For the purpose of estimating this property, an aerial with a gain of unity is regarded as an aerial which radiates equally in all directions —omni-directional—as if it were the centre of a sphere which is ever growing in size. If one considers the transmitter at the centre of an imaginary spherical balloon, then a receiver situated at any point on the surface would pick up the same amount of energy regardless of its position. However, as one gets further from the centre the signal strength reduces.

This form of radiation is wasteful because there are no television receivers in the celestial regions. Television aerials of the kind under consideration, therefore, are designed not to transmit spacewards (those for satellite applications, though, are different and are, indeed, designed for space beaming). When an aerial is designed to eliminate high-angle radiation the 'balloon' takes on the form of a doughnut, and if it is assumed that the transmitter power is the same as before, then a receiver located on the surface will pick up a signal of greater strength. The 'balloon' idea of aerial response is illustrated in *Figure 6.7*, where (a) shows the spherical concept with a gain of unity, and (b) the doughnut concept, where the gain is now increased to three times by virtue of this type of response characteristic.

This means that if the transmitter power is, say, 10 kW, then the aerial will make the *effective radiated power* 30 kW.

The response may be further modified, and such may be necessary to concentrate the signal energy over a largish angle, while eliminating or significantly reducing energy over a smaller angle at the rear as a means of avoiding interference on the receivers of a more distant service using the same channel numbers. Diagram (c) in *Figure 6.7* shows a response like this, which increases the gain to five times. Normally, stations of shared channel numbers are located as far as possible away from each other to avoid mutual interference, and further protection in this respect is provided by the rapid fall-off in signal field with increasing distance from a u.h.f. transmitter and, in some cases, by a change of

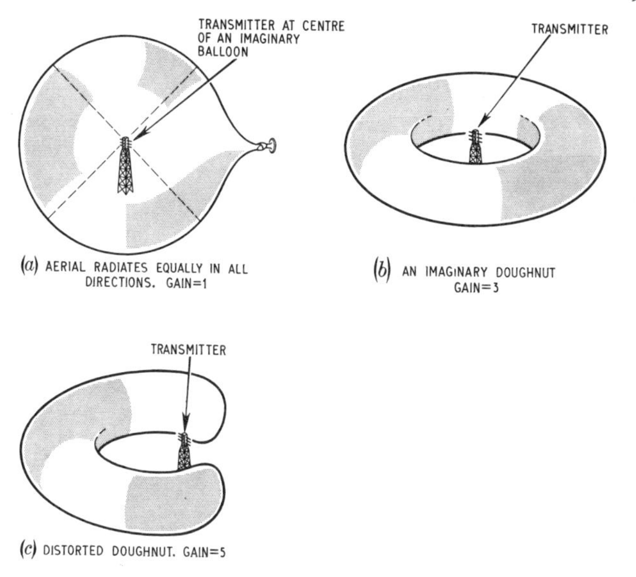

Figure 6.7. 'Balloon' concept of polar diagrams (see text)

polarisation. However, during spells of anticyclonic weather, u.h.f. signals can be propagated over far greater distances than is normal, and reducing the radiation from the aerial in the direction which could prove troublesome during reception conditions like this is another protection artifice.

The 'balloon' responses shown in *Figure 6.7* are called *polar diagrams*.

Vestigial sidebands

We have seen that the information transmitted by a modulated carrier wave lies in both the sidebands. Each sideband carries the same information and so they duplicate each other. It is possible to send a complete signal using one sideband only; and this can be desirable due to the consequent reduction in bandwidth. However, it is not easily possible to sever one sideband as cleanly as shown at (b) in *Figure 6.8*, so suppression usually takes the form of a 'tail' or a gradual roll-off of the unwanted portion. The choice is to have this occurring either in the wanted sideband or in the unwanted one, as shown by (c) and (d) in *Figure 6.8*.

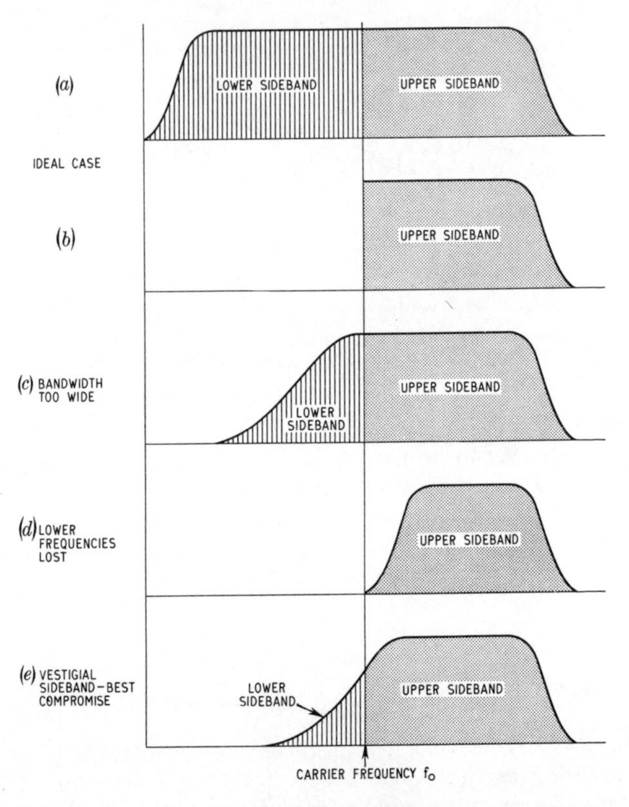

Figure 6.8. Sideband structure and vestigial sidebands. These are explained in the text

Neither of these alternatives is really satisfactory. In the first case low-frequency energy is lost if the roll-off occurs in the wanted sideband, while in the second case incomplete suppression occurs and the bandwidth of the signal as a whole is still fairly large. In television a compromise is sought, where the lower sideband is partially suppressed, and because there is a vestige of it remaining, the scheme is known as *vestigial sideband transmission*.

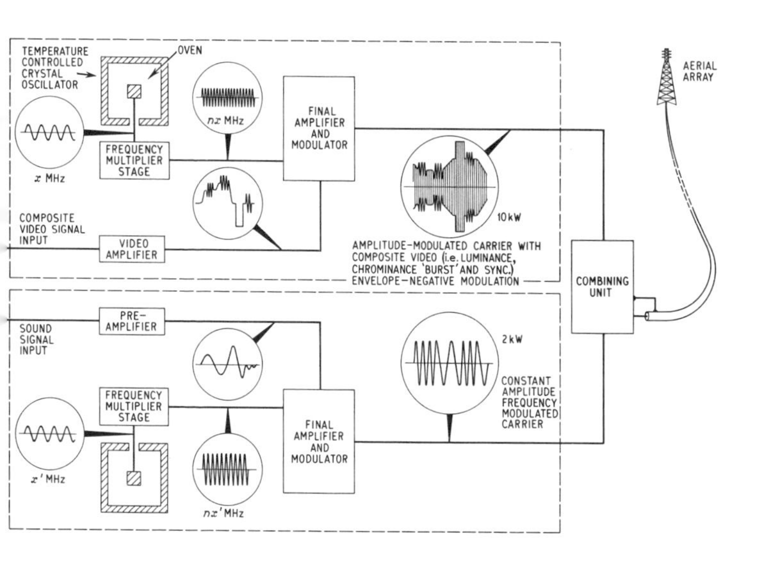

Figure 6.9. The prime elements of a colour television transmitting station

The net effect of this type of transmission when picked up by a receiver suitably adjusted to accept it is that the sideband energy is perfectly equalised, the reception being just the same as if both sidebands had been transmitted.

7

DISPLAYING THE COLOUR PICTURE

After its reception, the encoded signal is resolved into the three colour primaries again, and the display device in the receiver is arranged to handle them as if they were the three primary signals sent straight from the studio centre over three separate circuits. We shall be seeing later, in Chapter 9, how the receiver derives these signals; but right now we must investigate how the signals are translated into colour pictures.

One way to re-form the picture would be to adopt a set of dichroic mirrors in conjunction with three filtered monochrome picture tubes, working the opposite way to the mirrors and colour filters, etc., in the camera. For large-screen colour television projection systems this scheme is sometimes used, but it is not ideally suited to domestic applications. For this the more convenient colour picture tube is employed, and at the time of writing there are two versions in use. One is called the three-gun shadowmask tube and is so far the most popular. The other is a single-gun tube which produces three beams, called the Trinitron. The shadowmask tube was developed by the Radio Corporation of America and the Trinitron by the Sony Corporation of Japan. Let us look first at the shadowmask tube.

Shadowmask picture tube

The three-gun shadowmask tube is a development of the ordinary cathode-ray picture tube used in monochrome television receivers. The basic features of such a tube are illustrated in *Figure 7.1*. The electron gun at one end consists of a cathode sometimes in the form of a fine tube of nickel, coated with an oxide which is rich in free electrons, producing them in abundance when heated. Round the cathode is a

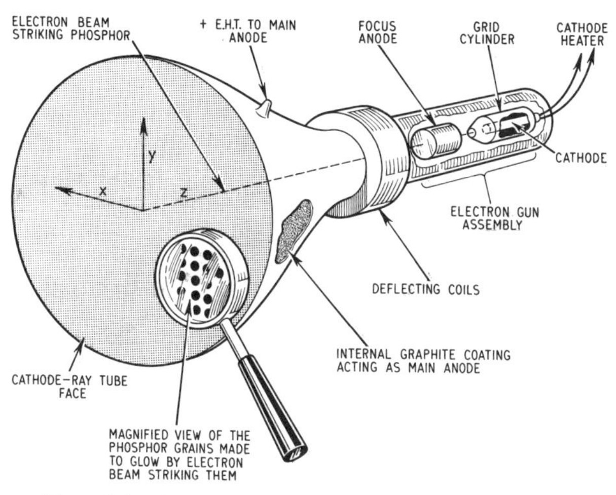

Figure 7.1. Basic principles of the cathode-ray picture tube

specially shaped cylinder which serves as the grid and controls the quantity of electrons leaving the hole at the end according to the negative potential, relative to the cathode, applied to it. The electrons that do leave are formed into a fine beam by another cylinder or a system of cylinders and/or other electrodes, depending on the gun design, so that the beam is focused and initially accelerated.

The essential reason why a beam of electrons actually leaves the gun assembly is because a final anode, round the inside of the flare of the envelope, is connected to a source of high potential 'positive electricity'. With shadowmask tubes the final anode is in the order of 25 000 volts and so the electrons, which are negatively charged particles, are accelerated towards it at very high velocity.

When the electrons strike the screen, the kinetic energy which they have picked up as the result of their mass in motion is exchanged for illumination, this being due to he electroluminescent nature of the phosphor material which is deposited on the inside of the screen. The electron beam thus produces a focused spot of light on the screen, and by a system of deflection this can be moved both vertically and horizontally, as indicated by the y and x axes in *Figure 7.1*. Moreover, by adjusting the negative potential on the grid the intensity of the beam and hence the spot brightness can be varied, giving another controlling 'dimension', signified z in the diagram.

The beam is deflected vertically and horizontally by the field and line timebases producing sawtooth current waveforms in the deflecting coils, this resulting in a changing magnetic field which deflects the beam in the appropriate directions. On the rising part of the waveform the spot is moved horizontally and less fast vertically, while on the fast falling part the spot very quickly returns to its starting point. The first part gives the forward trace and the second part the retrace or flyback as it is called. The actions yield a background of lines and it is upon these, owing to the brightness of the spot changing under the control of the picture signal, that the picture is constructed. It will be recalled from Chapter 4 that one set of lines is interlaced with a second set to give a picture of the total number of lines appropriate to the system. Thus we have two fields each composed of half the number of lines of a complete picture. The background of lines produces a rectangular area of illumination on the screen, and this is called the *raster*. Elementary aspects of scanning are given in Chapter 3, and for a more detailed treatment at beginner's level the companion book, *Beginner's Guide to Television* should be consulted.

If a magnifying glass of sufficient strength is focused onto the face of a monochrome picture tube, as shown in *Figure 7.1*, the surface will appear as a large number of tiny grains in random distribution. These are grains of the phosphor material. The term 'phosphor' is a generic one, used to describe this group of materials, and has no connection with the chemical term phosphor—as in phosphoric acid, etc.

Phosphors in the television sense have two important characteristics: the first is colour and the second glowing persistence. In television the persistence is desirably in the order of 100μs—just long enough to allow a line to remain before the next one begins with new picture information. In monochrome television the phosphor chosen is that which gives the most pleasing 'white' light.

The three-gun picture tube

Colour television picture tubes have to carry three different screen phosphors in order to produce three displays in

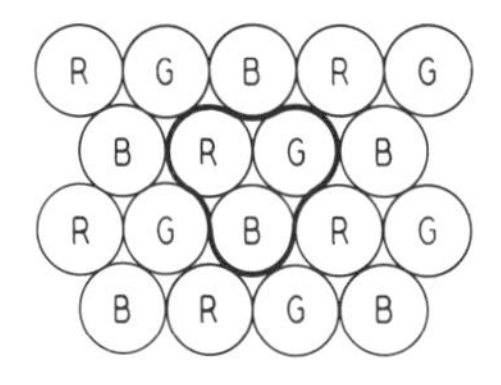

Figure 7.2. The formation of the red-, green- and blue-glowing phosphor dots on the screen of the three-gun shadowmask picture tube. The three-dot triangular formation is called a triad

perfect registration, one each in red, green and blue. If the face of such a tube is examined under a microscope or powerful magnifying glass the phosphor-dot formation might be seen, as shown in *Figure 7.2*. The phosphor make-up of the dots, of course, looks the same for each group of colour, and it is only when the phosphors are stimulated by the electron beams that the different colours are produced.

There thus exists a multiplicity of 'groups' of the three colour phosphors, and each group, detailed in *Figure 7.2*, is called a *triad* for obvious reasons. In a typical tube there are about 440 000 triads, all forming together an interwoven pattern over the entire screen area.

The three guns are arranged in the tube neck so that the electron beam from one energises only the red phosphors, that from the second only the green phosphors and that from the third only the blue phosphors. For this reason the guns and beams are often referred to as red, green and blue. They are not coloured of course! When a group of triads is caused to glow at proportioned colour intensities, the eye is deceived into concluding that the three primary colours occur at the same point and so it discerns a spot or picture element of a colour corresponding to the mix of the three lights. When the mix is correctly portioned for white light, the spot or picture element appears without colour as a 'shade' between peak white and black—e.g., a shade of grey.

Remembering what has already been said in the early chapters about additive mixing by lights, it is easy to see how the spot or picture element can have either of the three primary colours, red, green or blue, a complementary colour, yellow, cyan or magenta, or indeed, any colour of intermediate value as determined by the colour triangle or the chromaticity diagram (*Figure 5.2*). However, in practice the display hues are somewhat limited by the nature of the available primary colour phosphors. Nevertheless, the phosphors adopted by modern tubes make the display virtually subjectively equal in terms of colours to those reproducible by good colour photography.

The diagrams in *Figure 7.3* give a basic impression of the working of the shadowmask three-gun picture tube. It is, in fact, three tubes in one glass envelope and showing the same face. The three electron-gun assemblies work completely independently to produce the three separate electron beams, each of a different strength according to the colour and monochrome (e.g., luminance) signals by which they are being controlled.

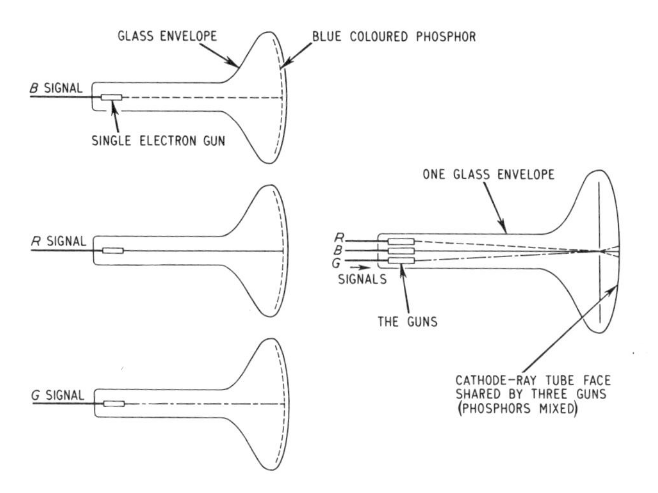

Figure 7.3. These diagrams show how the three-gun tube can be regarded as 'three-tubes-in-one'

Shadowmask

In addition to the components of the tube already mentioned, there is what is known as a shadowmask, from which the tube partly gets its name. In appearance this is something like a metal gauze carrying a large number of holes. Actually, there is one hole for each triad, and the purpose of the mask is to ensure that the phosphor dots of the different colours are hit only by the electron beams from the guns of corresponding colours. This will become apparent in a minute.

The shadowmask is located about 10 mm ($\frac{3}{8}$ in) behind the phosphor screen, and the three beams are caused to converge onto a single hole and pass through it to diverge slightly the other side so that the red, green and blue beams strike the red, green and blue glowing phosphor dots.

Each dot is then activated by the electroluminescent process, described earlier, to yield an amount of light proportional to the strength of the electron beam striking it. As already mentioned, the three dots so activated are so close physically that the eye is unaware of their separation, and so adds the three primaries together to produce the intended colour for that point, as illustrated in *Figure 7.4*.

The shadowmask is made of a thin sheet of metal about the size of the picture tube faceplate. It is manufactured by a photochemical etching process similar to that used for the production of lithographic half-tone plates for printing. Accuracy, however, is of extreme importance, so the tolerances acceptable are much smaller than those relating to printing plates.

Arranging the phosphor dots

To produce the phosphor-dot pattern a rather ingenious method is used. The green-glowing phosphor material is first sprayed onto the screen so that it covers the whole area, and then the shadowmask is mounted accurately in the faceplate. To cause the green phosphor material to adhere to the screen at the points corresponding to the dots, ultra-violet light is arranged to pass through the shadowmask holes at an angle corresponding to that of the electron beam from the green gun when the tube is correctly active. Account also has to be taken of the curvature of the electron beam in the region of the scanning fields, and for this an accurately positioned optical system is used for the exposure.

When the green phosphor has been selectively hardened by the ultra-violet light the shadowmask is removed and the screen is washed to remove the unexposed phosphor

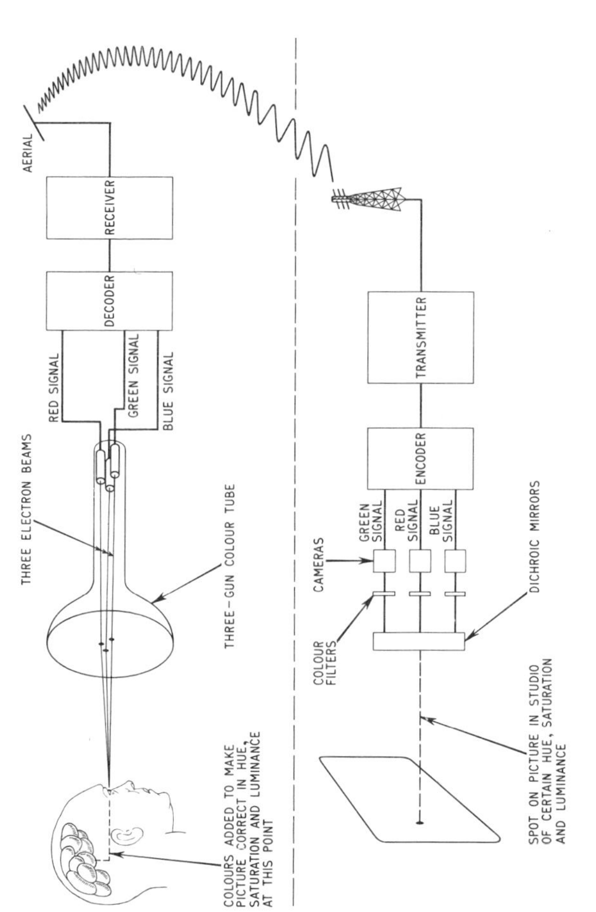

Figure 7.4. Instantaneous reception of a colour element using a three-gun tube

material. The screen then contains a uniform pattern of small dots composed of the green-glowing phosphor material.

This process is repeated separately for the red- and blue-glowing phosphor materials, with the ultra-violet light in these cases emanating from the positions which will be occupied by the red and blue guns.

Modern colour tubes have an aluminium screen backing, the same as monochrome tubes, which enhances the brightness.

The brightness and the nature of the white light produced by the tube are also dependent on the type of tube phosphors used. Various phosphors have been adopted over the years, and recent improvements have increased the general brightness of the white field at a beam current of 800 μA from an early value of 55 nit to something like 120 nit (the nit, sometimes nt, is the metric unit of luminance).

The primary colours are commonly identified in terms of x and y coordinates on the chromaticity diagram or colour triangle, and for academic interest, if nothing more, the coordinates of a 1970/71 shadowmask picture tube are: *red* $x = 0.655$ and $y = 0.347$; *green* $x = 0.330$ and $y = 0.605$; *blue* $x = 0.148$ and $y = 0.059$.

Other recent features relate to the pushthrough faceplate, so that the tube itself can also act as the implosion guard, the new aspect (width-to-height) ratio of the display of 4 : 3 from the earlier 5 : 4, compensation for the change in register of the shadowmask with temperature increase, etc.

The latter feature is particularly interesting when it is realised that a significant proportion (about 20 W) of the total beam power is collected by the shadowmask assembly, which then obviously heats up! This causes the dimensions of the shadowmask to alter with a consequent loss of alignment. The result of this would be a progressive loss of colour purity and white uniformity unless steps were taken to combat it. Modern tubes employ a bimetal support system which yields an axial movement of the shadowmask assembly to compensate for radial changes in register.

Purity and convergence

Figure 7.5 shows how the three beams pass through a hole in the shadowmask and then diverge slightly to strike the appropriate colour phosphors. This is achieved by the three guns in the tube neck being inclined inwards slightly, as shown by the top drawing.

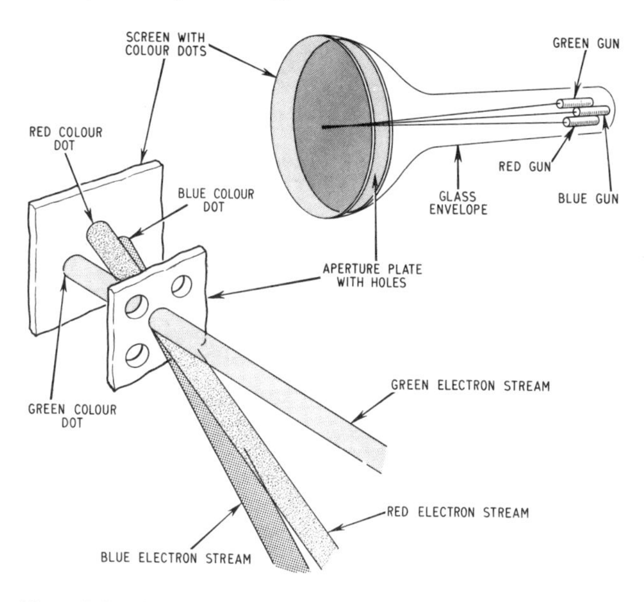

Figure 7.5. Basic principles of the three-gun shadowmask picture tube

However, this is not enough, and to compensate for manufacturing tolerances and variations in operating environment several adjustments have to be provided electronically and on the tube neck in order to regulate the beams and their approach angles, etc. There are two categories of adjustment, one is to ensure that each of the three beams strikes only its associated dots, so that with the green and blue beams switched off a *pure* red raster is obtained, and likewise green and blue rasters separately with the other two partnering beams inactive. However, colour *purity*, as this adjustment is called, is commonly attended to on a red raster with the green and blue beams inactive. If the red raster is contaminated with a 'sea' or blobs of different hues, then the tube is said to be 'impure' and a purity adjustment is required to correct the condition.

The other category of adjustment is *convergence*. This is necessary to ensure that all the picture elements register accurately on the red, green and blue displays, for it will be recalled that the complete picture results from the accurate interlaying of the three simultaneous displays in red, green and blue. If one display is out of registration with the other two, then the two in registration will fail to produce the correct colours (or greys) and the display in the colour which is out of registration will be displaced from the first.

For example, on a grey picture element (say a small square) blue misregistration would result in the display being in blue and also in yellow with either a vertical or horizontal displacement, depending on the nature of the convergence error. We would get the yellow display as well as the blue one because yellow is produced by the addition (by the eye) of red and green lights. When blue is added to yellow, of course, we get back to 'white' (or grey) again when the proportions of the three coloured lights are correct. Thus to correct for the error on the square display just mentioned the convergence would have to be adjusted to shift the displaced blue square onto the top of the yellow square, the result then being a grey square, assuming that there was no colour in the original scene or in the electronically produced signal.

Purity adjustment

A pair of small ring magnets provides for purity adjustment. The assembly is fitted onto the tube neck in partnership, so that the field intensity is varied when the two magnets are adjusted relative to each other, while the field orientation is varied when the two are adjusted (e.g., rotated) together. The magnetic field thus passes through the neck of the tube and deflects the three beams together, adjustment being made so that the angle of approach of the beams provides perfect 'on target' hits on the associated colour phosphor dots.

Convergence adjustment

Now, while correct purity adjustment ensures that the three beams strike the dots of corresponding colour, it fails to provide the conditions required for the three beams to pass through the same group of holes of the shadowmask. It is thus possible to have good purity and lack of convergence.

Misconvergence, for example, would result if the red and blue beams were to pass through the same group of holes while the green beam passed through the holes of an adjacent group. In this case the red and blue displays would be coincident (giving magenta) while the green display would be out of registration. It should be noted that while the bottom diagram in *Figure 7.5* gives the impression that the three beams together pass through only one hole at a time, in practice several holes are used, so that the effective scanning spot consists of a cluster of stimulated triads. However, it is easier to consider one hole and one triad in use at a time.

We must also remember that the three beams are being deflected vertically and horizontally together, so apart from the basic requirement of *static convergence* at the centre of the screen, there is also the very important requirement for the display to remain in correct registration (convergence) as the beams are fully deflected. This is called *dynamic convergence*.

As for purity, magnetic fields are used for convergence,

but while the field for purity is deliberately arranged to influence the three electron beams together, as we have seen, the fields for convergence are applied separately to each beam. The scheme is engineered in conjunction with the shadowmask picture tube and convergence assembly, located on the tube neck, so that the fields are concentrated round each beam with minimal interaction. Thus it is possible to adjust one of the beams for convergence without affecting the other two.

Static convergence is provided by a steady magnetic field either from a permanent magnet or an electromagnet (one for each beam), while dynamic convergence is provided by specially changing magnetic fields from electromagnets energised from signals in the line and field timebases.

The convergence assembly contains the components which produce the magnetic fields and in most of them there is an adjustable permanent magnet for each beam along with two electromagnets, consisting of two pairs of coils for each beam. One pair of coils receives the special signals from the line timebase, giving horizontal dynamic convergence for each beam, while the other pair receives the special signals from the field timebase, giving vertical dynamic convergence for each beam.

Figure 7.6 shows the three pairs of coils working in association with pole pieces located inside the neck of the tube, through which the beams pass. The dotted lines across the pole pieces indicate the lines of magnetic force, while the arrow-headed lines show the directions of beam deflection due to the magnetic fields. It should be noted that the pole pieces of the electromagnets line up with those in the tube neck and that magnetic shields isolate the three beams from each other.

When permanent magnets are used for static convergence, there is usually one for each of the three arms of the convergence assembly, and *Figure* 7.7 shows the arrangement for one of the arms, that corresponding to the blue beam. Here it is shown how the steady magnetic field in the pole

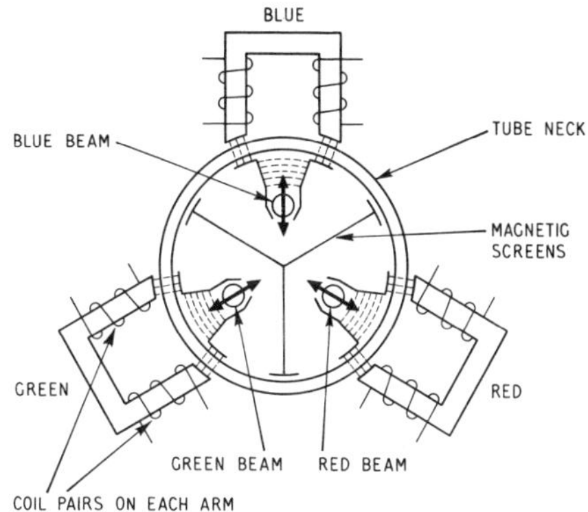

Figure 7.6. The three arms of the convergence assembly round the tube neck each relate to one of the beams, with the blue beam usually at the top. Each arm has a coil pair, one for the vertical dynamic convergence current and the other for the horizontal dynamic convergence current. The current in each coil is separately regulated in amplitude and waveform 'tilt' by the convergence controls

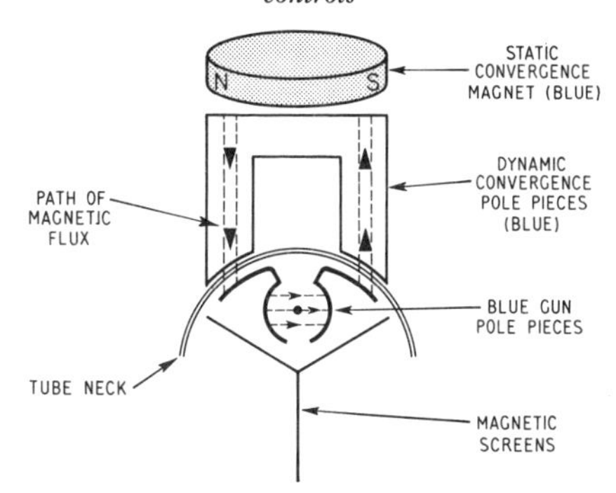

Figure 7.7. For static convergence each arm of the convergence assembly often embodies an adjustable permanent magnet as shown. Sometimes the steady magnetic field required for static convergence is provided by an adjustable d.c. in the dynamic windings

pieces can be adjusted by rotating the static convergence magnet for optimum static convergence at the centre of the picture.

In general, then, the magnetic fields cause radial shifts of the three beams so that registration is maintained both at the centre of the screen (static) and away from the centre as the beams are being deflected (dynamic). Since the electron beams can be regarded as conductors carrying an electric current, the deflection is thus at right angles to the magnetic field by an amount corresponding to its strength.

To optimise the convergence it is necessary to be able to shift the blue beam tangentially as well as radially as shown in *Figure 7.8*. Note that the blue beam is at the top with the

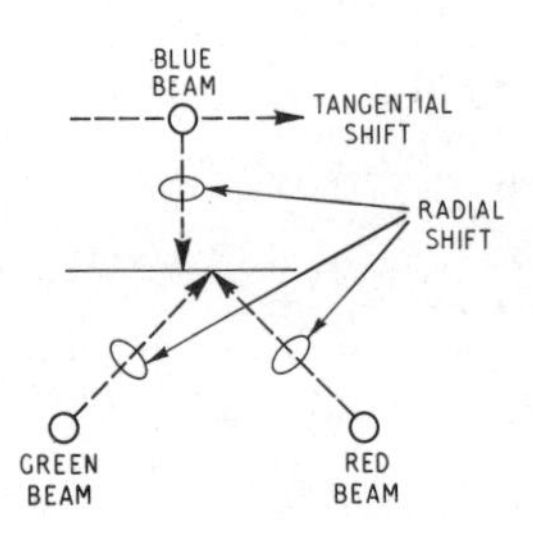

Figure 7.8. Showing the radial and tangential shifts required for convergence. A tangential shift is provided on the blue beam only, in addition to radial shifts on the three beams

red and green beams on either side in delta formation. To provide the tangential shift a blue lateral assembly is also located on the tube neck. This usually embodies an adjustable permanent magnet which, in conjunction with its pole pieces, produces radial lines of force and hence a lateral or tangential shift of the blue beam only. Sometimes the assembly also carries an electromagnet which is energised with signal from the line timebase.

While it is not too difficult to appreciate the function of static convergence, that of dynamic convergence is somewhat more complex. As the three beams swing across the face of the tube the convergence requirements change for each beam for two main reasons. One because of the displaced electron guns in the neck, the three beams thus not being on a common axis, and two because the radius of deflection of the three beams differs from the radius of the shadowmask and screen—the trend nowadays being to secure the flattest screen or faceplate which, of course, is incompatible with the radius of deflection.

Without correction the red, green and blue rasters would resolve separately as shown in *Figure 7.9*. The dynamic

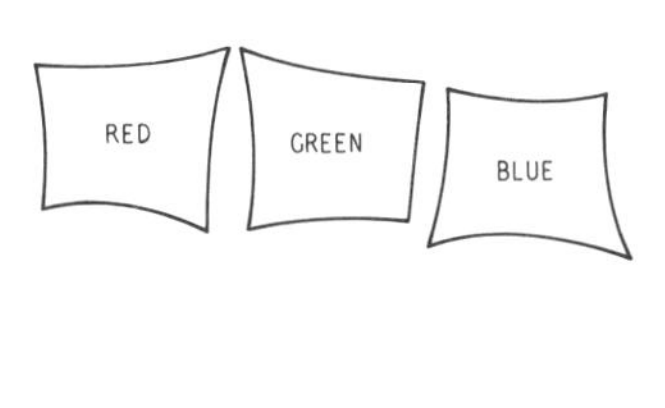

Figure 7.9. Owing to the displaced electron guns in the shadowmask tube and the difference between the radius of deflection and the radius of the screen, raster distortion of the kinds shown occurs. It is the job of the convergence systems to correct these distortions so that the three pictures in red, green and blue appear on top of each other in the best possible registration. The pin-cushion residual distortion remaining on the compounded pictures is often cleared by a transductor working between the field and line scanning currents

convergence fields, derived from the timebase signals, are therefore very carefully tailored to improve the form of each raster so that the residual distortion of each is then the same. When this happens the three correspond accurately on top of each other and the elements of each picture register over the entire screen area. In practice slight misconvergence can be detected, particularly at the corners of the screen, but the error diminishes (or should do) into insignificance at the normal viewing distance.

One electromagnet of a pair receives signal from the line timebase while the other of the same pair receives signal from the field timebase. It can be shown mathematically that the signal waveforms need to be of a parabolic nature. However, to optimise the adjustments, which are usually carried out with the screen displaying a dot or crosshatch pattern, the waveform tilt and amplitude have to be regulated for each beam in both the vertical (field) and horizontal (line) directions. An array of dynamic convergence controls is provided for this purpose and the service manual details the exact procedure for adjusting them.

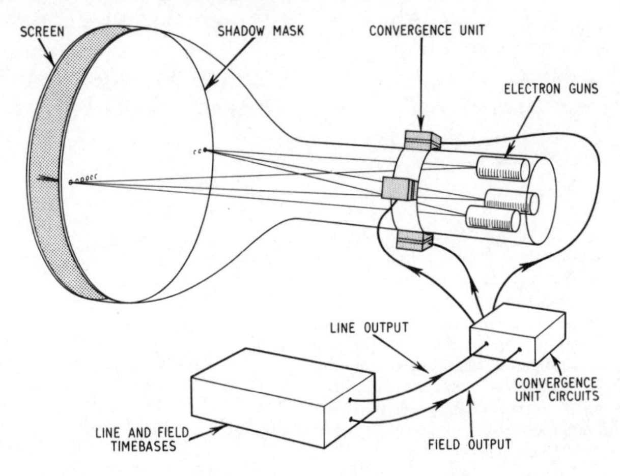

Figure 7.10. The basic principle of dynamic convergence

It is noteworthy that the residual compounded raster distortion is essentially of pin-cushion shape, and this is cleared by a special *transductor* which operates between the line and field scanning waveforms. For a more detailed examination of convergence the reader is referred to my book entitled *Colour Television Servicing*, by the same publishers.

To summarise, the diagram in *Figure 7.10* gives the elementary concept of convergence.

Grey-scale tracking

As already noted, the final anode of a three-gun shadowmask tube requires up to 25 kV for optimum performance. Beam focusing is achieved by an electrostatic lens system in each gun and this works from a potential between about 4.2 and 5 kV. The first anodes (or screens) operate between about 210 and 495 V relative to the corresponding cathodes, while the grids swing from about −135 V for beam cut-off to −65 V for maximum brightness.

Now, a very important requirement is for the beam current/grid voltage characteristics of the three guns to match over the full range of brightness. Since it is impossible for guns to be manufactured with this parameter exactly matching, the tube has to be equipped with various adjustments to provide accurate *grey-scale tracking*, as it is called.

We have seen that white light is produced when the mix of red, green and blue lights is correctly proportioned. This must occur at all settings of the brightness control, for a drift of one gun or more out of alignment as the control is adjusted would alter the intensity of one colour or more and thus change the proportions of the mixture, which would result in the white (or grey) display becoming coloured. The effect on a colour scene would be to alter the hue or saturation with changing luminance level. The beam current/grid voltage characteristics are matched by adjusting the voltages of the first anodes at lowlights and the luminance drive to the guns at highlights, using a stepped contrast display. The idea is to secure the grey steps between black and peak white without colour contamination.

Need for degaussing

We have seen that some adjustments of the shadowmask picture tube are provided by controlled magnetic fields. This means, therefore, that external fields could interfere with the display if they were allowed to affect the three beams, thereby putting the convergence and purity—particularly the latter—in error.

To reduce the effects of extraneous fields, including the

Earth's magnetic field, a part of the tube's flare is equipped with a magnetic shield. It is essential, therefore, to ensure that this and other nearby ferrous metal items are free from residual field before making any adjustments to the purity and convergence. The device employed for this is the *external degausser*, which is a coil composed of a large number of turns of insulated copper wire energised from the 50-Hz (or 60-Hz) mains supply. The coil is generally wound for about 1000 ampere-turns from the 240-V supply. It is held parallel to the tube face and moved slowly away in circular motions. Similar treatment might be required by the metal parts at the rear and round the sides of the tube, but the field should never be allowed to influence the magnetic components on the tube neck.

Modern receivers are less affected by extraneous fields than earlier models whose tubes had less effective magnetic shielding. At one time it was necessary to degauss, as just described, each time the position of the receiver was altered, owing to the resulting change in the Earth's magnetic field relative to the tube. All modern receivers are equipped with an internal degausser which operates each time the receiver is switched on. There are usually two series-connected coils in proximity to the tube's magnetic shield which pass a fairly heavy alternating current from the mains supply initially on switching on, this then diminishing swiftly as the control circuits take effect, so that negligible current flows when the receiver is working normally.

The Trinitron colour tube

This tube is already in use in Sony receivers in various parts of the world, including the U.K. It differs significantly from the shadowmask tube in that there is only a single gun. However, this produces three beams which can be independently modulated by the three primary colour signals. The three cathode sections of the gun are in a common horizontal plane, so the three beams are produced in a horizontal line. Instead of the phosphor-dot triads of the shadowmask tube, the screen of the Trinitron is composed of several hundred vertical stripes of red-, green- and blue-glowing phosphors, and instead of a shadowmask, there is a metal aperture grille which has a slot for each three (red-, green- and blue-glowing) phosphor stripes. The general principles of the design are illustrated in *Figure 7.11*.

The vertical stripes, of course, are very close together so that at normal viewing distance the three colours at any

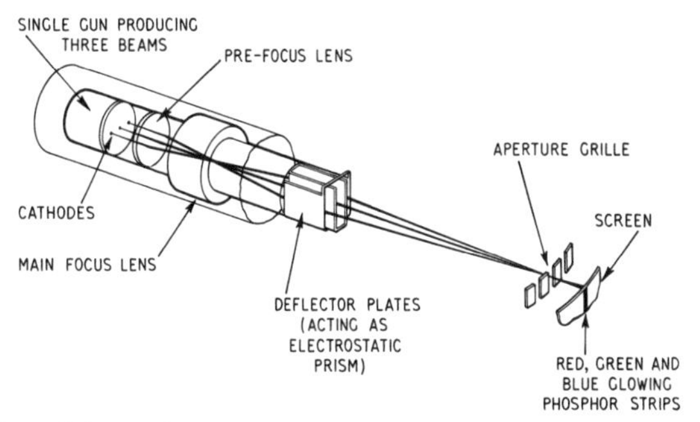

Figure 7.11. The main features of the Trinitron single-gun, three-beam aperture-grille picture tube

instant merge to give the impression of the correct colour of that particular element. The function in this respect is similar to the shadowmask tube.

The tube has several desirable features: the aperture grille has a greater beam transparency than the shadowmask so that a display of given brightness can be obtained with less e.h.t. voltage, which in effect means a brighter picture; the extraneous magnetic field influence is less than with the shadowmask tube; convergence is simplified since the three beams are on a common axis, it thus being necessary merely to arrange for some means of adjusting the angles of approach of the two outside beams, so that they converge at the aperture grille with the beam emanating from the middle cathode, an adjustment which in practice is provided by the electrostatic prism shown in *Figure 7.11*; and the tube is less likely to produce the Moiré pattern which is sometimes seen on shadowmask tubes (particularly early ones) owing to the interference between the hole structure of the shadowmask and the line structure of the raster.

The Trinitron which is used in some of the Sony colour sets requires 19 kV e.h.t., a focusing electrode potential from zero to 400 V (significantly less than the shadowmask) and a second grid potential of 240–450 V. It would seem that the Trinitron tube is likely to have a greater scope of application in the future since at the time of writing Sony was considering licensing tube manufacturers to make it and also selling the tube to set manufacturers. Developments along these lines are also taking place elsewhere.

8

DOMESTIC AERIAL SYSTEMS

Although it is not possible to see the radio wave which links the transmitting and receiving aerials, quite a lot is known about it and diagrams can be constructed to express its behaviour.

The electrical 'disturbance' which occurs in proximity to the transmitting aerial may spread outwards equally in all directions, getting weaker all the time as the distance from the transmitter increases. The field strength within the horizon distance of the waves used for television broadcasting is proportional to the square root of the radiated power and inversely proportional to the square of the distance. The signal wavelength and the heights of the transmitting and receiving aerials also come into account in absolute terms.

An aerial that radiates in all directions is termed *isotropic*. In practical terms, however, this is not possible, since even in the simplest case the Earth's surface interferes with the energy pattern, and as has already been explained (Chapter 6) a television transmitting aerial is deliberately made directional.

Radio waves possess a number of important parameters. One is the plane of polarisation which can be particularly important in television.

Magnetic and electric fields

Any radio wave consists of two energy-carrying fields—magnetic and electric—which together form the electromagnetic wave (which is the radio wave, light waves, alpha waves, etc.). The two fields are at right angles to each other and both are at right angles to the direction of propagation,

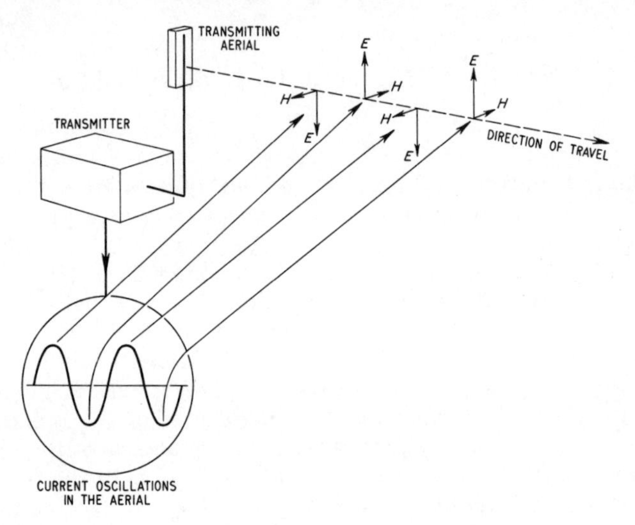

Figure 8.1. Vectorial concept of an electromagnetic wave

as shown in *Figure 8.1*. The information carried by the waves resides in the increase and decrease of the fields with time as they travel along.

Plane of polarisation

The plane of polarisation of the wave is defined by convention as the orientation of the electric field. If one half-cycle of the waveform in *Figure 8.1* is regarded as positive, then the arrow representing the electric field could be drawn pointing upwards, and when a negative half-cycle occurs it will point downwards.

The electric field will always act along a line which may be drawn parallel or vertical to the ground. Thus, when a communications engineer speaks of the plane of polarisation of an electromagnetic wave he means that the electric field is either parallel or vertical to the ground, corresponding to a horizontal or a vertical plane of polarisation. These terms are illustrated in *Figure 8.2*, and from them it will be appreciated that the plane of polarisation could fall at some

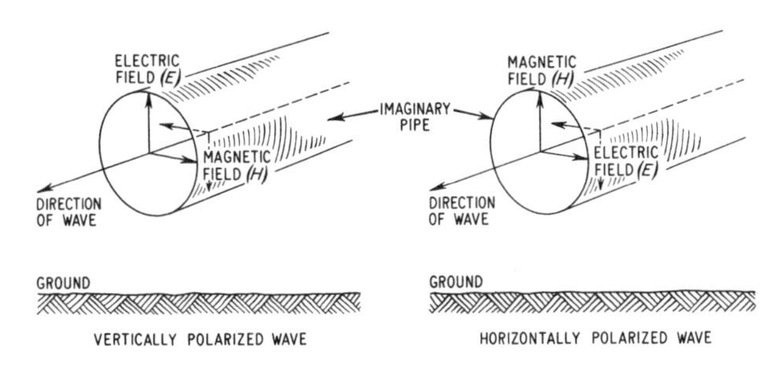

Figure 8.2. Illustrating the plane of polarisation of an electromagnetic wave

angle between the vertical and horizontal. In television, though, either vertical or horizontal polarisation is adopted (it is noteworthy that 'slant' polarisation is used at some local v.h.f. broadcasting stations to facilitate signal pick-up on vertical car-mounted rod aerials without detracting too much from the pick-up on horizontally mounted domestic aerials). Horizontal polarisation is employed on the v.h.f./f.m. systems of sound broadcasting.

The plane of polarisation of the signal is determined initially by the transmitting aerial, and the receiving aerial should normally be orientated in the same way for the maximum signal energy to be abstracted from the passing electromagnetic wave. The same principle applies to radiated polarised light which, as already noted, is also an electromagnetic transmission. For example, if light of a given plane of polarisation is radiated from a suitable source, then at the receiving point, which could be, say, a photoelectric cell, a filter or lens of corresponding polarisation is required for maximum response.

Giving a transmission vertical polarisation in one area and horizontal polarisation in another area significantly helps to reduce interference under abnormal reception conditions when the two transmissions have the same frequency (e.g., co-channel working) because a vertically mounted aerial will

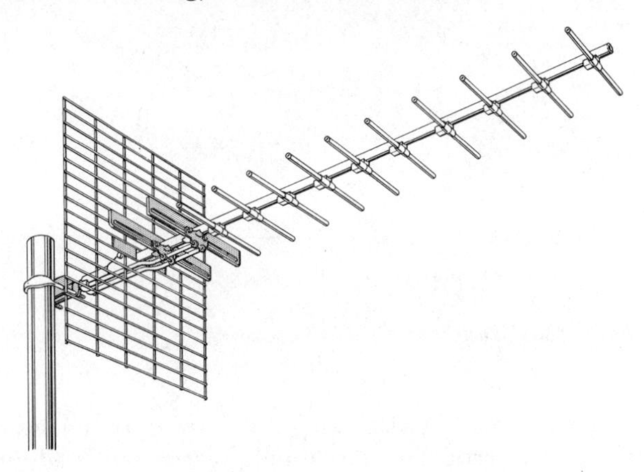

Figure 8.3. U.h.f. Yagi array set for receiving a signal of horizontal polarisation

respond only very slightly to a horizontally polarised signal, and vice versa. At the time of writing most u.h.f. transmitters adopt horizontal polarisation, and an aerial suitable for such a transmission is shown in *Figure 8.3*. However, there are some u.h.f. transmitters scheduled for vertical polarisation, and a few of these are on the air.

Wavelength and frequency

Two other important parameters of a radio wave are frequency and wavelength. These are inter-related by the velocity at which the wave travels such that frequency (f) $\times$ wavelength (λ) = velocity (v).

Since in free space the velocity is constant at 300×10^6 metres per second, it follows from the expression that the higher the frequency the lower the wavelength. Thus to find the wavelength in metres we merely have to divide 300 by the frequency in MHz. This is because 1 MHz is equal to 10^6Hz or cycles per second, which is an old term. Similarly, we can find the frequency in MHz by dividing 300 by the wavelength in metres.

Bandwidth

Aerial elements are usually referred to in terms of wavelength, and an aerial is usually said to be 'tuned' when its length corresponds to one-half of the wavelength of the signal it is to receive. Actually, there is a small difference between the aerial half-wavelength and the free-space signal half-wavelength. This is because the velocity reduces slightly when the wave is 'accepted' by the aerial. The physical half-wavelength of the aerial is thus about 95 per cent of the signal half-wavelength.

As the wavelength of the transmission reduces, so the tuned aerial half-wavelength gets smaller, and at u.h.f. the active element may not be much more than about 200 mm (8 in). However, to avoid clipping the wanted sidebands of the colour signals the aerial must have a sufficient bandwidth to pass not only the signals of one channel, but also those of the other two (later three) channels in the local group. As noted in Chapter 6, the local group of channels may extend over 88 MHz or more. If the aerial bandwidth is insufficient or if the aerial is of the incorrect channel grouping for the area the chroma sidebands may be seriously attenuated. This will either suppress the colour altogether or result in severe desaturation. In some cases the luminance and sound signals may also be affected, and one channel of a local group might be received better than the other two. Such imbalance effects often signify the use of the wrong aerial for the location or an aerial of poor design. The effects, though, can be aggravated by local screening and 'standing wave' conditions.

Aerial size

The monochrome 405-line system used to (and still does for the time being) work in the v.h.f. spectrum, so a single-channel aerial might be between 4 m (11 ft) and 1 m (3 ft) in length, depending on the local channel number(s). *Figure 8.4* shows at (a) a single-element aerial for the Band I 405-line system which, for a 50-MHz signal would be about 3 m (10 ft) in length, and at (b) the simplest form of single-element u.h.f. aerial, being between about 177 and 330 mm in length. The first is for a vertically polarised signal and the second for a horizontally polarised signal.

A single element constitutes the *dipole*, and a signal from this is obtained from terminals at the centre and is passed to the receiver through feeder cable, which in the U.K. is commonly coaxial cable having a flexible outer sheath and an inner conductor supported through the centre of the sheath by low-loss insulating material, called the dielectric. This has a characteristic impedance, as it is called, of about 75 ohms. The receiver's aerial input circuit is designed to match this impedance, while the impedance at the centre of the dipole is of a like value. The half-wave dipole is usually in two half-sections with an insulator at the centre to provide mechanical continuity. The outer conductor of the coaxial cable is connected to the inner end of one half-section and the inner conductor to the inner end of the other half-section, at the insulator.

The single dipole as shown at (b) in *Figure 8.4* might possibly work reasonably well close to a u.h.f. transmitter where there is little shielding and the signal field is high. However, when the receiver is some distance from the transmitter (say, 20 to 30 miles) or when it is near a source of

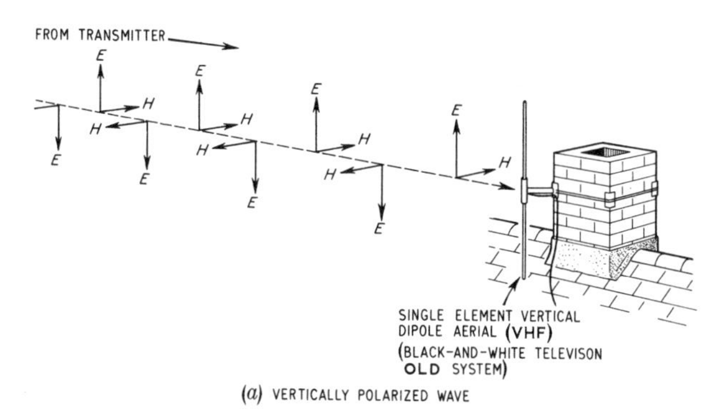

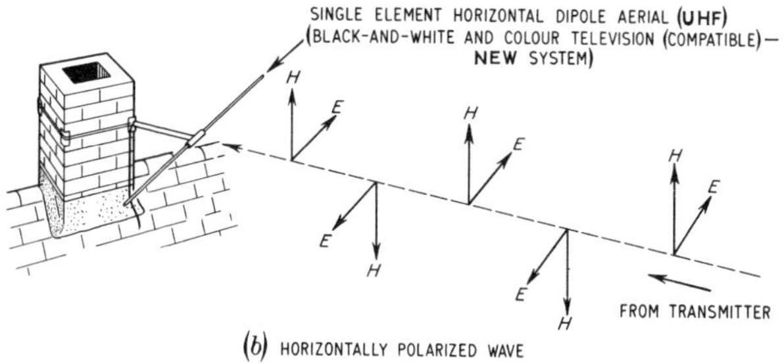

Figure 8.4. V.h.f. aerial set for receiving a signal of vertical polarisation (a) *and u.h.f. aerial set for receiving a signal of horizontal polarisation* (b). *Both aerials are single centre-fed dipoles*

electrical interference, then the single dipole alone will be insufficient to give a high enough signal-to-noise ratio. To increase the directivity and hence to 'focus' the receiver onto the transmitter more aerial elements will be required.

Polar diagrams

This term is used when discussing the directivity of aerial arrays, and it is necessary to understand at least the significance of the term, which has already been partly explained in Chapter 6, when we were dealing with transmitting aerials.

The polar diagram is a graphical way of showing on paper the directional properties of an aerial array in one plane only. For television this is taken as looking down on top of the array; that is, in a plane horizontal to the Earth which is regarded as a 'flat' surface.

There are two ways of plotting polar diagrams for any aerial system. In the first the array is miniaturised (e.g., accurately scaled down) so that it can be placed in the centre of a laboratory at a given height (say, 3 m, 10 ft). This technique is illustrated in *Figure 8.5*, which also shows measurements being made on the 'model' array.

The main reason for scaling down is to make the array more manageable both from a mechanical and an electrical point of view. Because the scaling down is accurately performed, the polar diagrams obtained will correspond equally to the full-scale array.

Some example polar diagrams for different arrays are given in *Figure 8.6*. These were obtained by using the array under test as the transmitting aerial and moving a sensitive receiver, equipped with a suitable readout device, round the perimeter at given angles some distance from the array. The amount of signal pick-up is then plotted at each angle and the diagram completed by joining the points together.

The same directional properties apply when the array is used for reception. For example, the same plotting would be obtained by taking a transmitter round the same perimeter and measuring the pick-up on a calibrated receiver attached to the array under examination.

If point A on the first polar diagram in *Figure 8.6* is facing the transmitter all the available signal will be picked up and

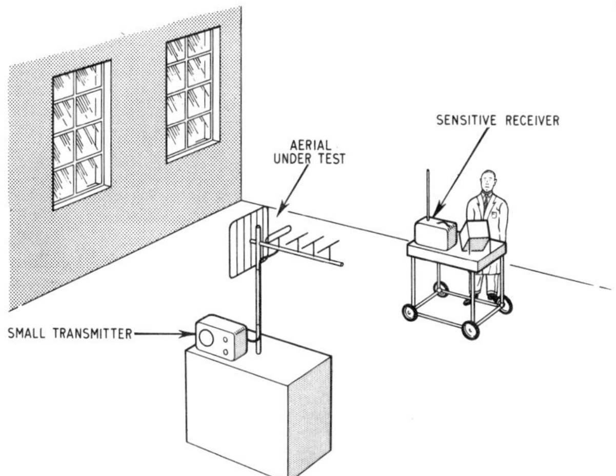

Figure 8.5. By using the test aerial for transmission and receiving the radiated signal at various angles round it on a calibrated receiver, a polar diagram can be plotted as explained in the text

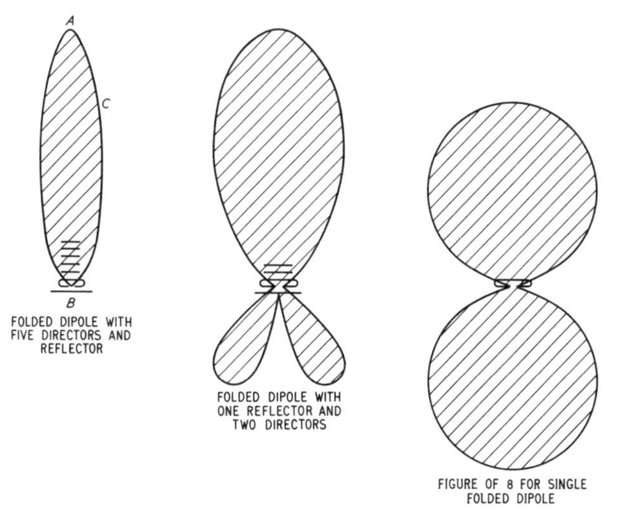

Figure 8.6. Example polar diagrams (see text)

a noise source at point B will fail to interfere with the signal. It is sometimes necessary to have point C looking towards the transmitter instead of point A, so that the zero pick-up point B can be aimed at the noise source. Although this will not yield the maximum signal available, it will exclude or considerably reduce the interference and thus result in an improved S/N ratio, which is the aim at all times.

Yagi array

The most popular type of television aerial to date is the Yagi or 'end-fire' array as it is sometimes called. This was developed as early as 1928 and has been much employed for short-wave activities. During the Second World War it was adopted for early radar applications where a small angle of acceptance was important. It consists basically of a half-wave dipole, one or more reflectors and a host of directors.

When the array is used for reception the signal energy is 'guided' into the dipole which, remember, is the active element, by the directors and any that gets past is returned to the dipole by the reflector(s). The net result is an array of high gain (compared with a single dipole) and a narrow polar diagram.

Whether an element is to act as a reflector or director depends on its length relative to the dipole. When it is longer it behaves as a reflector and when shorter as a director. The spacing between the various elements is also of importance in determining the behaviour of an array.

Indeed, there is much more to the design of a good u.h.f. wideband Yagi array than may meet the eye. It has to satisfy such parameters of gain, polar diagram, front-to-back ratio (indicating the degree of rear pick-up to front pick-up), bandwidth, terminal impedance over the bandwidth (ideally this should remain resistive at the nominal designed-for impedance, which is usually 75 ohms in the U.K.), etc.

The gain tends to increase as more directors are added, but the law of diminishing returns eventually comes into play, and unless special care is taken over element lengths and spacings a side lobe may spring up on the polar diagram, which would make the array responsive to interference arriving broadside on. U.h.f. arrays usually have a reflector composed of several rod-type elements or of a 'mesh' construction, as seen in *Figure 8.3*.

Folded dipole

The reflector and director elements are so-called 'parasitic', meaning that they are not normally connected electrically

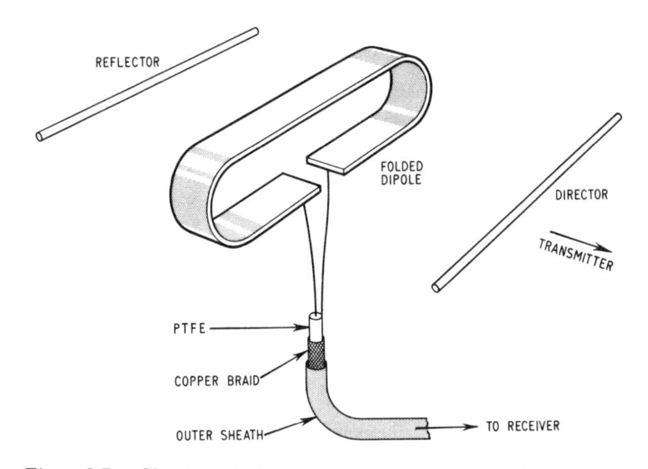

Figure 8.7. Showing a folded dipole connected to a coaxial cable feeder. Notice the reflector behind and the director in front, relative to the direction of the transmitter

to the feeder or dipole. As more such elements are added, so the centre impedance of the dipole falls and in order to maintain a reasonable match to the feeder the dipole may have to be folded as shown in *Figure 8.7*. This artifice steps up the centre impedance of the dipole by four times and thus compensates for a four-times reduction from the single-dipole centre impedance of 75 ohms, as might well happen due to the presence of a reflector system and directors.

Other matching schemes have been adopted over the years and some of these relate to the maintenance of the impedance over the full bandwidth; thus they should never be altered.

The Yagi array has undergone various degrees of modification since its inception for u.h.f. television. Different manufacturers have introduced different styles and types of

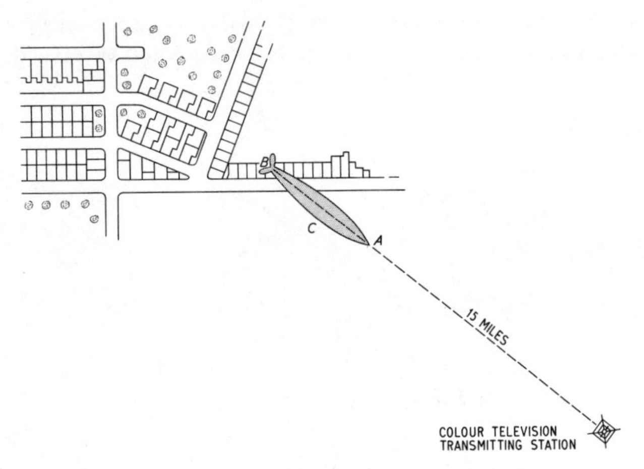

Figure 8.8 Polar diagram of Yagi array orientated for the best signal pick-up

parasitic elements, and two or more *pairs* of Yagi arrays are sometimes coupled together in special ways to enhance the gain and to provide polar diagrams of specific requirements.

To summarise this chapter, *Figure 8.8* shows the orientation in terms of the polar diagram, of a high-gain u.h.f. Yagi array so as to receive the maximum signal from the transmitter. Remember, the aerial can be the weakest link in the chain, and if this is poor the picture will never be good no matter how fine the receiver!

9

THE PAL RECEIVER

A colour receiver can be regarded as a monochrome receiver with refinements to which has been added the colour display device, the circuits required by the display device and the colour decoder. *Figure 9.1* shows the block diagram of a colour receiver with the extra items required for colour in heavy line.

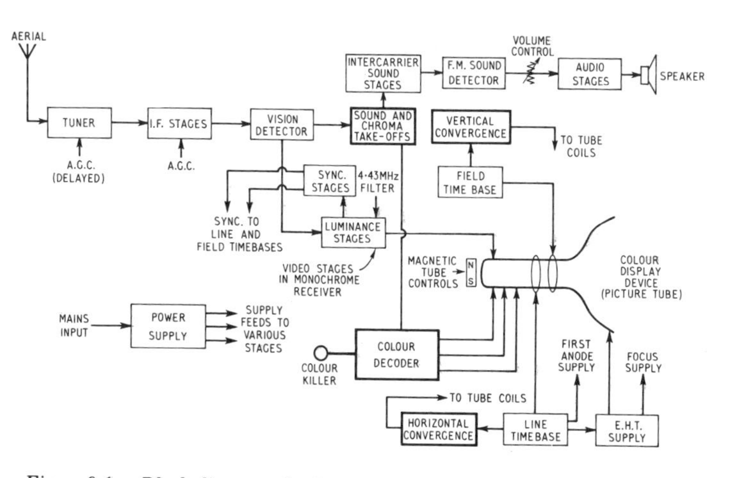

Figure 9.1. Block diagram of colour receiver. The heavy-line blocks represent the extra stages required for colour

The colour display device is shown in heavy line because this is specific to a colour receiver, but one should not conclude from this that a colour receiver is unable to display colour in monochrome or, indeed, fail to reproduce in monochrome from a non-colour transmission. We have seen from Chapter 7 that colour display devices demand control circuits and potentials in addition to those from which an ordinary monochrome picture tube can operate. These are not related directly to the circuits and stages through which the signals pass. This, then, leaves the main and essential item of the colour receiver, which is the decoder.

The sound and chroma take-off 'block' is also shown in heavy line. This is merely because some receivers have special circuits in this area to assist with the separation of the chroma information, the luminance signal and the intercarrier sound signal. From first principles it would be possible to separate the three signals at the vision detector and from there send them on their separate journeys.

What we are endeavouring to highlight is the fact that by removing the colour decoder most colour receivers would operate perfectly normally on a monochrome signal—in black and white of course! Under such conditions it would also be possible to substitute a monochrome picture tube for the colour display device (with appropriate adjustment to the electrode potentials) and get the receiver to operate essentially as a monochrome one.

The colour decoder, in fact, incorporates an automatic switch, called the colour killer, which makes it functional only when the aerial signal contains colour information. It 'demutes' the decoder by responding to the colour bursts which an encoded signal carries. Thus, when there are no bursts the decoder remains quiescent and the colouring items of the receiver, excepting the tube and its controls and potentials, fail to contribute to the display.

The companion volume, *Beginner's Guide to Television*, traces the various circuit sections and describes their functions in some detail, but for the sake of continuity it will now be desirable briefly to cover some of this ground again relative to the items in *Figure 9.1* which are common to both monochrome and colour, then concentrating afterwards on the items which actually add the colour to the monochrome picture. At this juncture it is noteworthy that a receiver designed for any colour system contains in essence, though with some variations in detail, all the stages shown in *Figure 9.1*.

The tuner

The tuner's job is to amplify the weak signals supplied by the aerial, to select or 'tune' the required one, and then to translate this, together with all the information that it carries, to a lower frequency, called the intermediate frequency (i.f.). It will be recalled that the aerial is a tuned 'circuit', so the aerial and the tuner should really be regarded as an integrated 'packet'. If this way of looking at the front-end was appreciated more by some aerial erectors and installation technicians quite a few more viewers would seemingly be getting better pictures than they are at present, particularly in difficult reception areas.

It should be remembered that the selected signals are in a specific channel and consist of both the modulated sound and vision carriers. Also bear in mind that the aerial must be designed to deliver to the tuner equally the sound and vision signals of the three local group of channels. Imbalance in this respect tends to encourage poorer reception on one channel relative to that on the other channels, the poor reception often being revealed by excessive grain (due to tuner noise) on the display. Similarly, imbalance over any one channel upsets the sound/vision ratio, while also sometimes affecting the colour display by emphasising or attenuating the colour subcarrier relative to the main vision carrier. The receiver can accommodate such imbalances to some extent by virtue of the automatic gain control (a.g.c.) in the main i.f. channel and the automatic colour control (a.c.c.) in the chroma channel, about which more anon.

The tuner has three stages, which are the r.f. amplifier, the mixer and the local oscillator, as shown by the block

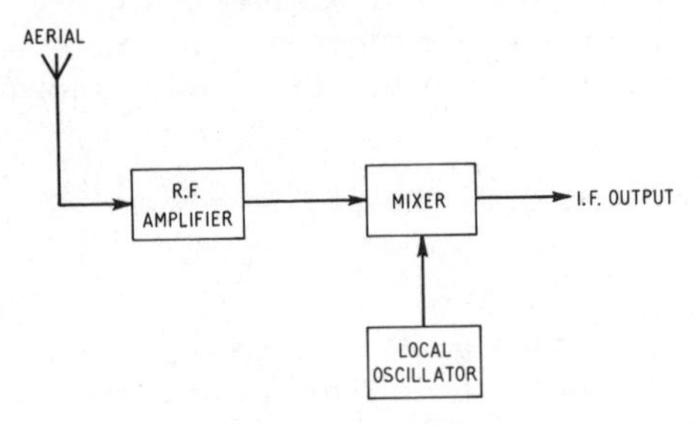

Figure 9.2. Block diagram of tuner

diagram in ***Figure 9.2***. The mixer thus receives amplified aerial signals and the local oscillator signal, and the design of this stage is such that sound and vision signals corresponding to the frequency difference referred to the oscillator signal are produced at the output. The mixing process, of course, yields sum and difference frequencies, but the local oscillator

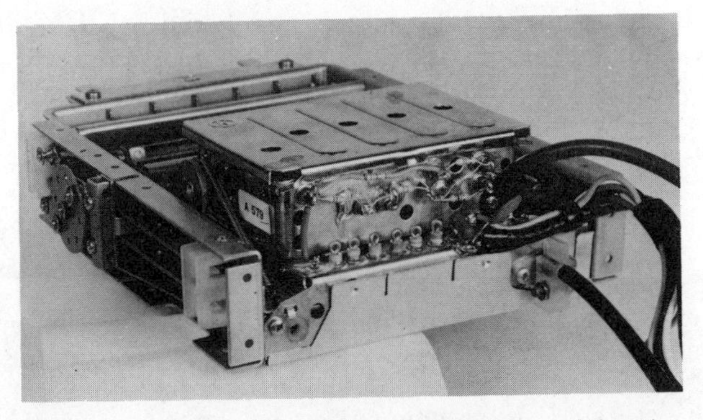

Figure 9.3. Tuner used in a Bush receiver containing a u.h.f. section based on ganged tuning capacitors

frequency is arranged so that the difference frequencies correspond to the i.f.

The displacement between the frequencies delivered by the r.f. amplifier and the local oscillator must be maintained at all selected channels, and this is achieved by the channel selection altering the tuning of the r.f. amplifier, mixer input and local oscillator in step, so that the i.f. difference is always maintained.

Figure 9.4. Philips receiver employing the 'G8' chassis (shown hinged out) which features a capacitor-diode tuner. The presets for this can be seen in the bottom left-hand corner, next to the loudspeaker

Some u.h.f. tuners incorporate ganged variable capacitors mechanically coupled to the tuning knob or station selector buttons, while the latest ones use so-called 'electronic tuning' where, instead of mechanical capacitors in a ganged assembly, capacitor-diodes are employed. These are essentially p-n semiconductor junctions in reverse bias which, under this condition, exhibit capacitance between their lead-out wires, the value decreasing as the reverse bias is increased (this happens because of the widening of the depletion layer, which constitutes the dielectric between the two 'plates' formed by the heavily doped p and n regions). Thus merely by arranging for the tuning control or selector buttons to alter the biasing and hence the capacitance the frequencies corresponding to the required channels can be tuned.

An example tuner incorporating a u.h.f. capacitor-gang section is given in *Figure 9.3*, while *Figure 9.4* shows the rear of a Philips receiver containing the 'G8' chassis which embodies a capacitor-diode-type u.h.f. tuner. The presets for this can be seen in the bottom left-hand corner, next to the loudspeaker.

To achieve the required degree of front-end selectivity, most u.h.f. tuners employed in receivers destined for the British market are equipped with a total of four tuned circuits (which are quarter-wave lines), one for the local oscillator and three in front of the mixer for tuning the r.f. amplifier and a bandpass-coupled circuit between the r.f. amplifier and mixer. This is generally necessary in the U.K. to avoid interference due to 'image' (second channel) responses from other channels operating in the u.h.f. bands. European countries currently appear to have less trouble in this respect and some u.h.f. tuners operating in receivers on the Continent may incorporate only two tuned circuits prior to the mixer.

The mixer/local oscillator function is called 'frequency changing' for obvious reasons, and sometimes just one stage is used for this. The stage concerned is then often known as a 'self oscillating mixer'. All recent receivers employ low-noise, u.h.f. transistors in the tuner.

I.f. stages

Looking again at *Figure 9.1*, the i.f. signals (sound and vision) from the tuner are amplified by the i.f. stages, which now use transistors. The response characteristics of the stages are carefully tailored to suit the sound and vision, including the subcarrier, signals and various filter circuits are present here.

The idea of the i.f. channel, therefore, is not only to step up the relatively weak tuner signals to a level suitable for operating the vision detector, but also to adjust the relative amplitudes of the signals of the tuned channel and to introduce filtering to remove unwanted signals of adjacent channels. Thus both amplification and selectivity are provided by the i.f. channel circuits.

Vision a.g.c.

Automatic gain control is often applied both to the tuner r.f. amplifier and the i.f. channel. This is an automatic control whereby the gain of the stages concerned are caused to increase when the signal strength falls and decrease when it rises, thereby ensuring that the vision detector receives a relatively constant level signal input irrespective of the channel tuned and the prevailing signal conditions.

The control potential for this is obtained either by sampling the black level intervals of the video signal or the peak amplitude of the line sync pulses. The sampling is commonly at line timebase repetition frequency, and a rectifier circuit and filter translates the sampling action to a d.c. potential which is then fed as a bias to the controlled stages. So-called 'forward a.g.c.' is commonly applied to n-p-n transistor stages, such that the gain of the stage receiving the potential decreases as the potential or bias increases.

It will be understood, of course, that the rise and fall of the control potential is directly geared to changes in received signal strength.

If the gain of the tuner r.f. amplifier is reduced too early the signal-to-noise ratio tends to suffer and the picture displays 'grain' interference. This is avoided by introducing a delay to the a.g.c. potential applied to the r.f. amplifier, such that the gain of the i.f. channel is first reduced, after which further control is applied to the r.f. amplifier, with rising signal strength.

Vision detector

The correctly tailored and normalised i.f. signal thus arrives at the vision detector, and it is the job of this stage to extract the information originally modulated onto the carrier wave (see *Figure 9.5*). The vision detector output, therefore,

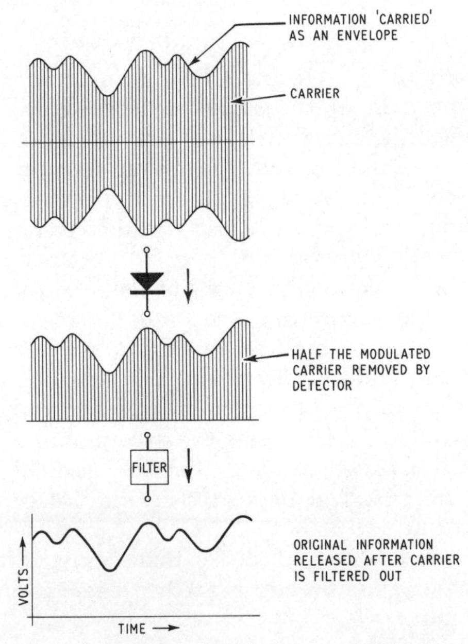

Figure 9.5. Illustrating the basic principles of a.m. detection

consists of the complex modulation carrying both luminance and chroma information. Filtering is employed to remove residual i.f. signal.

Vision detector output

In a monochrome receiver the vision detector output goes direct to the video amplifier and thence to the picture tube cathode. In a colour receiver, however, there is some difference here because the chroma signal is also present at the vision detector output and this has to be separately channelled to the chroma stages in the decoder. To avoid confusion, therefore, we will first run through the various stages which are active both on monochrome and colour, and afterwards examine in detail the decoder and colour tube control circuits. We can thus assume until then that our example colour receiver is responding only to a monochrome input, with the colour killer actively muting the decoder.

Intercarrier sound

Since at the detector there are both the sound and vision signals (since both together were handled by the i.f. channel with suitable level control), and because the detector is a non-linear device, intermodulation of the two signals occurs such that a third output at a frequency corresponding to the difference between the vision and sound carriers is generated. On the 625-line U.K. system the sound/vision spacing is 6 MHz, so the third signal is at 6 MHz. Owing to the manner in which it is produced it is referred to as the inter-carrier signal. It will obviously possess information corres-ponding both to the amplitude modulation of the vision signal and to the frequency modulation of the sound signal, but at this point we are interested only in the f.m. content, and this can be extracted merely by passing it to an f.m. detector, which responds minimally to the a.m. signal. Since it is the sound information which we need, therefore, the 6-MHz signal is usually referred to as *intercarrier sound*.

A filter at the vision detector or at a stage following the detector tunes out the 6-MHz intercarrier sound channel, which might also contain amplitude limiting to help with the suppression of the vision a.m. components. Again, this channel is generally transistorised.

The amplified intercarrier sound signal at the output of the channel is then applied to an f.m. detector, which may be the normal ratio detector (using a pair of diodes) or a different arrangement based on an integrated circuit, the latter assuming greater popularity with the progression of solid-state electronics.

This yields the audio signal, which then goes to the audio amplifier, via the volume control, to the sound output stage and thence to the loudspeaker.

Luminance channel

The video signal also present at the detector is fed to the luminance or Y amplifier channel which, incidentally, is the direct equivalent of the video amplifier in a monochrome-only receiver. However, while in a monochrome-only receiver only a single valve may be used, in a colour receiver there are several stages, the latest receivers using all transistors and some of the older ones transistor pre-amplifiers followed by a valve output stage.

The luminance channel also incorporates the brightness control and sometimes the contrast control circuitry, the former for adjusting the bias, via the luminance channel d.c. couplings, of the picture tube guns and the latter for adjusting the level of the luminance signal proper.

It will be seen from *Figure 9.1* that the output of the luminance channel is coupled direct to the picture tube, often to the cathodes of the three guns, via preset adjustments for establishing the correct grey-scale tracking (in conjunction with presets arranged to regulate the tube first anode potentials).

We shall be seeing soon that while the bandwidth of the luminance channel must be sufficiently wide to provide full picture detail (around 5.5 MHz), that of the chroma channel is little more than ±1 MHz because a good colour picture can be obtained with lower colour definition provided the background luminance is itself resolved in full detail.

Now, when a signal is passed simultaneously through two channels, one with a wider bandwidth than the other, it tends to undergo a slight delay in the narrower bandwidth channel. For this reason, therefore, a delay line of about 600 ns is included in the luminance channel circuit to delay the signal in this channel so that its components will arrive at the picture tube at exactly the same moment in time as the corresponding colour signal components fed to the tube through the reduced bandwidth chroma channel in the decoding section.

The required luminance channel bandwidth is obtained by special circuit design, sometimes assisted by h.f. compensating circuits, and the channel will often be found to contain a retrace blanking circuit, whereby during the line and field retraces of the timebases the picture tube guns are heavily biased off to ensure that the retrace (or flyback) lines are removed from the display.

The a.g.c. potential is also derived in the luminance channel, the video signal at black level or at the tips of the line sync pulses being gated in some way at line frequency. Various circuits have been involved for a.g.c.

Sync separator

The luminance (e.g., video) signal also feeds the sync separator, which is a relatively simple circuit designed for removing the picture information, leaving only the sync pulses, as shown in *Figure 9.6*. The line and field sync pulses are then separated and fed to the appropriate timebases to keep them in perfect step with the line and field scans at the transmitter (see Chapter 3).

It is worth noting here that some receivers derive the a.g.c. potential from the sync separator and also, in some

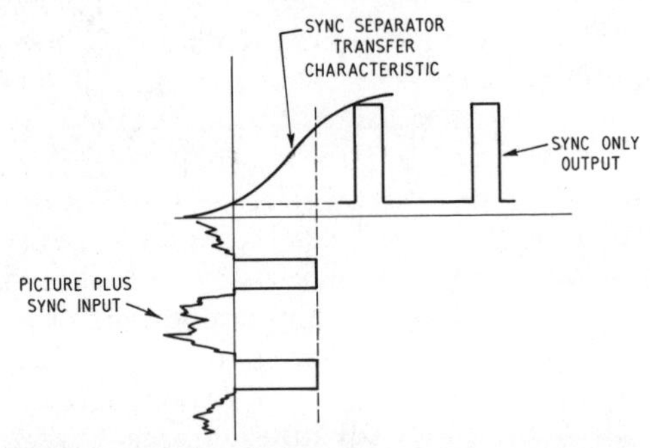

Figure 9.6. Illustrating the basic function of the sync separator. In this case the circuit responds only to the sync pulses, removing the picture information. The line and field sync pulses are then channelled to the corresponding timebases. Modern receivers use transistor sync stages

cases, the potential for operating the contrast control, this latter then being used to regulate the bias of the i.f. stages, and hence the i.f. channel gain, as the contrast control is rotated.

Remember, too, that a delayed proportion of the a.g.c. bias is usually applied to the tuner, and there might also be a manual gain control.

Timebases

The vertical and horizontal scans are produced respectively by the field and line timebases subjecting the electron beams in the picture tube to appropriate deflecting forces (by electromagnetic means), as explained in Chapters 3 and 7.

A substantial amount of magnetic force is needed to provide full deflection of the beams vertically and horizontally, and this force emanates from the scanning coils on the neck of the tube, the deflecting fields passing through the neck and thus influencing the electron beams. *Figure 9.7* illustrates a colour television deflection unit by Mullard.

Figure 9.7. Mullard deflection unit type AT1014

The line timebase also drives a special circuit for producing the required 25 kV of final anode potential for the picture tube. This may be a separate e.h.t. generator, driven from the line pulses, or a voltage tripler unit, which receives pulses from a winding on the line output transformer at a level of about 8 kV. The latter is commonly adopted in contemporary receivers.

There is also some form of e.h.t. current limiting to prevent damage to the tube (or overheating of the shadowmask) on abnormally bright pictures or should the brightness control inadvertently be turned up too high. The control often comes from the line output stage and acts upon the brightness control circuit in a manner to reduce the beam current above a preset e.h.t. current level.

The e.h.t. tripler nowadays incorporates a subsidiary feed for the tube focus electrodes, which is adjustable to allow the focusing to be optimised. Colour picture tubes do not use the magnetic focusing of monochrome counterparts since the magnetic field would influence both purity and convergence (see Chapter 7).

Retrace energy in the line timebase is exploited to produce the required potential for the picture tube first anodes. These are connected to the supply via grey-scale tracking presets, which work in conjunction with the presets feeding the luminance signal to the picture tube.

Dynamic convergence

The line and field timebases feed convergence circuits which generate the current waveforms required by the dynamic convergence coils on the tube neck.

The various adjustments associated with the dynamic convergence are detailed in Chapter 7, in which the need for convergence is also explained.

Static convergence and purity

Static convergence and purity of the shadowmask picture tube are handled by adjustable permanent magnets on the tube neck, and these are represented by the 'block' showing

N and S poles in *Figure 9.1*. Again, Chapter 7 explains the need for these.

Power supply

The majority of colour receivers work from the domestic mains supply (some are designed to work from batteries), and the power supply in the receiver is designed to translate the 240 V 50 Hz mains supply to suitable d.c. voltages for operating the valves and transistors. There is also a transformer which steps down the a.c. to a value suitable for the tube (and valves when used) heaters.

In transistor receivers in particular regulator circuits are included to keep the feed voltages constant over the dynamic working range of the various circuits. It is easier with transistors to include the regulation here than, for example, in the line timebase circuit. The circuits are also fused and some are equipped with overvoltage or overcurrent cut-outs, which may need resetting manually or which reset automatically.

It is very important to ensure that any input mains voltage adjustment is set accurately to suit the local mains supply voltage (nominally 240 V 50 Hz in the U.K.).

Assuming that the tube electrodes connected to the decoding section are properly 'clamped', then the circuits so far described are those required for a monochrome display.

Colour decoding section

We are now left with the colour decoding section and the various stages and circuits therein. Quite a lot of background information on encoding/decoding has already been given in Chapter 5, dealing with the PAL signal, and it would be as well to run through that chapter again if necessary before going any further.

Chroma channel

The chroma channel is contained in the block diagram in *Figure 9.8*. It consists of the two chroma amplifier stages, with the a.c.c. and colour killer connections, the delay line driver, the PAL delay line and the PAL matrix. The other sections, although being associated with the chroma channel, will be considered later.

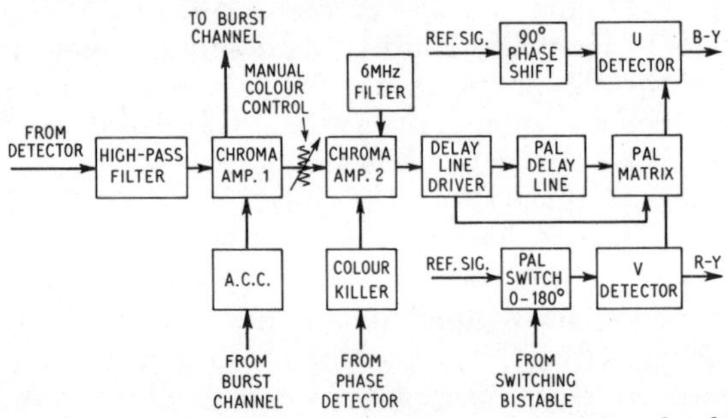

Figure 9.8. Block diagram of chroma section of colour decoder. Each section is fully explained in the text

Now, the composite signal from the vision detector, this time assuming that the receiver is correctly responding to a colour signal, is passed to chroma amplifier stage 1 through a high-pass filter. Since the chroma information is centred on 4.43 MHz, the high-pass filter accepts all this information while attenuating the lower-order components constituting the luminance information. Thus in essence only chroma information arrives at the input of the first chroma stage. The signal is then amplified by the second chroma stage and by the delay line driver, the amplifier section overall being engineered to provide a bandpass characteristic of ± 1 MHz centred on 4.43 MHz. Of course, there cannot be a sudden cut-off at 3.43 MHz and 5.45 MHz, but the skirts of the response are fairly rapid, the roll-off being aided at the h.f. side by the 6-MHz filter and at the l.f. side, as we have seen, by the input high-pass filter. Tuned circuits within the amplifier section may also be used to tailor the bandpass characteristic, sometimes shaping the top to minimise certain types of colour distortion.

Manual control of the level of the chroma signal is provided by the colour control, while automatic control (a.c.c.) is also provided at the first chroma stage. The second stage in the block diagram is controlled by the colour killer, but there may be differences in disposition of these controls in different receivers, though the basic principles remain unchanged.

The a.c.c. is generally applied to the stage as a potential which regulates its biasing and hence gain depending on the level of the chroma signal, as reflected by the amplitude of the colour bursts, which are picked up from the burst channel and then rectified to yield the control potential.

The second chroma stage in this illustration is deliberately designed to be non-conducting (e.g., biased-off) in the presence of monochroma input. This allows the receiver to work in black and white without the colouring circuits being active. If the colouring circuits were active during this time colour interference and noise might mar the monochrome display due to spurious signals getting to the picture tube through the decoding section.

Now, when the colour killer detects a colour transmission by the presence of bursts it produces a bias which automatically opens up the chroma channel, thereby letting the colouring signals reach the picture tube.

The colour killer generally receives ripple signal (at half line frequency—see Chapter 5—from the phase detector), and this it, or a part of it, rectifies to produce the switch-on bias for the controlled chroma stage.

The delay line driver thus delivers normalised chroma signal both to the PAL delay line and to the PAL matrix, and it is this terminating end of the chroma channel which deletes the 'phase sensitivity' of the chroma signal, such that the V and U components are neatly separated and fed to their appropriate detectors.

The PAL function

We have already dealt with this PAL function from the signal point of view in Chapter 5, but since it is particularly important no apology is made for harking back to it again,

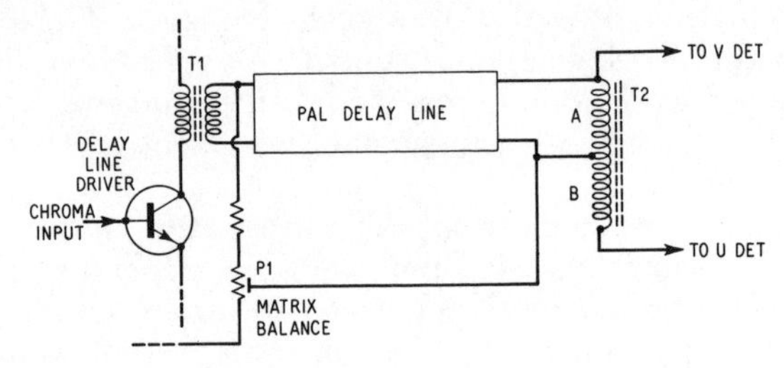

Figure 9.9. PAL delay line and matrix of a colour receiver. The action of this circuit is explained in the text

but this time more specifically from the point of view of the receiver.

The relevant circuit section of the PAL delay line and matrix is given in *Figure 9.9.* Some sets employ an arrangement just like this, while others differ in detail. Nevertheless, the net result is just the same whatever the circuit arrangement.

Chroma signal is fed into the delay line driver transistor base and appears amplified across the primary of transformer T1, whose secondary couples it to the delay line. Chroma signal is also picked up from preset P1 and from this fed to the output of the line. The output transformer T2 thus receives two lots of chroma signal, one delayed through the line and the other direct. It is the job of T2 to add and subtract (e.g., matrix) direct and delayed lines of chroma signal. These processes, it will be recalled, remove the 'phase modulation' from the chroma signal.

It works like this. We get addition and subtraction because the transformer is bifilar wound and centre-tapped. The effect is that the signal from the delay line across winding

A is exactly 180 degrees out of phase with the signal also across that winding from preset P1, while the signal from the delay line induced into winding B is exactly in phase with the signal also across that winding from preset P1. The antiphase signals across winding A thus cancel out, assuming equal strengths (one is subtracted from the other), while the in-phase signals across winding B add together.

Let us first consider the action on the U components of the direct and delayed signals. The subcarrier upon which the U signal is based is not phase reversed during alternate lines at the transmitter, so it has the same base phase on every line. Since the delay (equal to 63.934 μs, which is equivalent to the time taken by 283.5 cycles of reference signal—subcarrier) is exactly equal to one line period, we get from the adding function $U+U = 2U$, and from the subtracting function zero $U-U = 0$, assuming equal strength direct and delayed signals of unity value. The end of bottom winding B on T2 in which the addition occurs is thus connected direct to the U chroma detector, which then receives only the U chroma signal and no V chroma signal.

The V components of the direct and delayed signals are processed similarly, but the net result is different because (see Chapter 5) the transmitted subcarrier upon which the V chroma is based is reversed in phase during alternate lines, which means that the V components themselves are similarly processed. The subtracting part of the matrix thus 'sees' V signal of, say, 'positive phase' via the delay line and V signal of 'negative phase' via the direct route (bearing in mind the one line delay). The action on one line period is thus $(+V)-(-V) = +2V$, and on the next line period $(-V)-(+V) = -2V$. The subtraction occurs in winding A of T2, so the top end of this is communicated direct to the V chroma detector. This output, of course, carries no U chroma signal because this is exactly cancelled out due to the subtraction on each line.

To summarise, therefore, the U detector receives only U chroma signal because the alternate line phase reversals of the V chroma signal result in cancellation of the V chroma components in the adding section, while the V detector receives only V chroma signal because the antiphase V components are actually added in the subtracting section while the in-phase U components are subtracted. In other words, the composite chroma signal is very neatly separated into its two V and U components, the components now carrying information only in terms of amplitude *and not phase*. This means that phase distortion (phase distortion is tantamount to 'timing errors') in the system no longer introduces colour errors on the display. If system phase errors are fairly large there is a mild by-product effect of reducing saturation, but this is by far more subjectively tolerable than changes in hue!

We still need to introduce the reference signals to the V and U chroma detectors in phase quadrature (e.g., 90-degree difference in phase), of course, to simulate the quadrature modulation of the subcarrier at the transmitter. We also need to neutralise the effect of the $\pm$V chroma signal at the V detector. In *Figure 9.9* P1 adjusts the matrix balance so that the direct and delayed signals have equal strength. Imbalance here can encourage Hanover bar interference (see Chapter 5).

We can now return to *Figure 9.8* again and see that the reference signal to the U detector undergoes a 90-degree phase shift which provides the quadrature requirement just noted, while the reference signal to the V detector passes through a 180-degree PAL switch to neutralise the phase reversals of the V signal. On one line the relative phase of the reference signal to the V detector is say 'zero' and on the next line it is switched by 180 degrees, on the next line 'zero', the next 180 degrees and so on. The switch thus operates effectively at line timebase repetition frequency, but remember that the phase of the reference signal at the V detector is of the same value on every other line.

The PAL switch commonly consists of a pair of diodes switched alternately to phase reversing windings of a transformer passing the reference signal to the V detector, the actual switching being done by a bistable generator (this is of the multivibrator family but is stable in both switching modes and requires a switching pulse to trigger it from one mode to the other) triggered by pulses from the line timebase. It thus switches mode line by line, each mode having a switching frequency equal to half the line timebase repetition frequency. One diode is switched by the transistor in one half of the circuit, while the other diode is switched by the transistor in the other half of the circuit, there being two transistors in a bistable.

Provided the switching count corresponds to the plus and minus phases of the V chroma signal, therefore, the V detector will work as though it is receiving a V chroma signal of constant base phase; but if the switching count is in error the V detector will fail to work correctly and the displayed hues will be in error. To ensure that the switching count is correctly synchronised to the V chroma phase reversals line identification pulses are also fed to the bistable from the phase detector responding to the swinging bursts, but this story is yet to come.

To conclude the section of the decoder given in *Figure 9.8* it should be mentioned that some receivers employ the PAL switch between the PAL matrix V output and the V chroma detector. The effect is just the same as switching the phase of the reference signal to the V detector.

Burst and reference signal channels

Another section of the PAL decoder is given by the block diagram in *Figure 9.10*. Here the burst gate and amplifier receives composite chroma signal from the chroma channel, and filtering is included at 4043 MHz, the subcarrier frequency. The gate part is a transistor which is switched at line frequency such that it conducts only during the periods of the bursts. During the lines of picture information it is non-conducting. This means, then, that only the bursts of the chroma signal are amplified and that the output from this 'block' consists only of a series of bursts with no picture content.

Note that the bursts are also fed to the a.c.c. circuit in *Figure 9.8* which, after rectification and suitable processing, provide the control bias for the chroma amplifier, already considered.

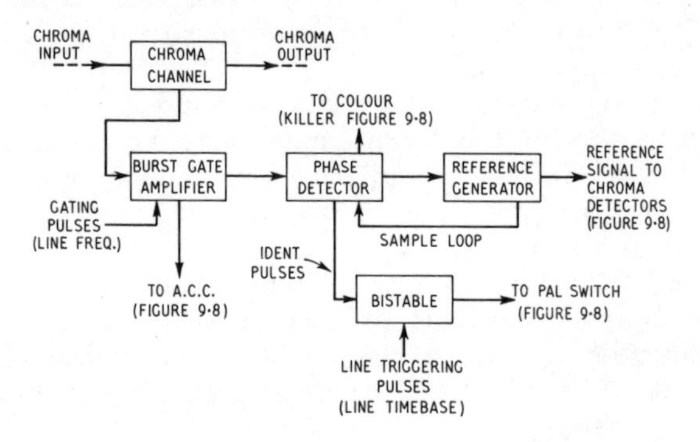

Figure 9.10. Block diagram of reference and burst channels

In the main, however, the bursts are fed to a phase detector which compares their average phase (remembering that they are alterating themselves in phase ± 45 degrees relative to the $-$U chroma axis, see Chapter 5) with the phase of the reference signal generated by the reference generator. The reference signal for such phase comparison is fed via a 'sample loop'.

The reference generator is crystal controlled but mild phase variation of the reference signal is made possible by a capacitor-diode effectively in shunt with the crystal. We have already come up against the capacitor-diode in electronically tuned u.h.f. tuners, and it will be remembered that such a device exhibits a decreasing capacitance across its lead-out wires when the reverse voltage is increased due to the widening of the depletion layer.

Thus it is possible to change the phase of the reference signal merely by changing the bias across the capacitor-diode. Now, for correct chroma detection it is essential for the phase of the reference signal applied to the detectors to correspond very closely to the phase of the subcarrier which was suppressed at the transmitter. The 90-degree phase shift required between the two detectors is provided by the phase shifter in *Figure 9.8*, which has already been mentioned.

The job of the phase detector

It is the job of the phase detector to provide a control potential for the capacitor-diode such that the phase of the reference signal is always held constant. This is easily possible for the average phase of the bursts corresponds exactly to the average phase of the subcarrier at the transmitter. The phase detector, in fact, produces a d.c. output of a positive or negative value corresponding to the phase error between the bursts and the reference signal. This is superimposed upon the standing reverse bias applied to the capacitor-diode so that the phase of the reference signal is continuously held at the correct value, with tendencies towards error being immediately checked.

The correctly phased reference signal is then fed to the V and U chroma detectors via the sections shown in *Figure 9.8*. Some receivers employ a d.c. amplifier between the phase detector and the capacitor-diode, while there is always a 'buffer' stage between the reference generator and the chroma detector feeds.

Ident pulses

Now, because the bursts are swinging in phase line by line such that on one line the phase relative to the −U chroma axis is +45 degrees and on the next line −45 degrees, etc., these swings being geared to the phase reversals of the V chroma signal, a facility is provided for the identification of the +V and −V lines (odd and even lines, Chapter 5) of chroma signal, thereby allowing the PAL switch to be correctly synchronised.

So far as the phase detector is concerned the swings of phase are processed as phase modulation, which means that the phase detector yields output pulses at half line frequency (about 7.8 kHz)—at *half* line frequency because the phase is the same on every other line.

These pulses, called 'ident pulses', short for identification pulses, are fed to the bistable, and the circuits concerned are arranged so that when the bistable is switching on the correct count for ±V chroma synchronising they have no influence on the normal triggering by the line timebase pulses. However, if at receiver switch-on the ±V chroma synchronising is in error, then the ident pulse at that instant will oppose the line timebase triggering. The bistable will miss one count and thereafter count in correct synchronism.

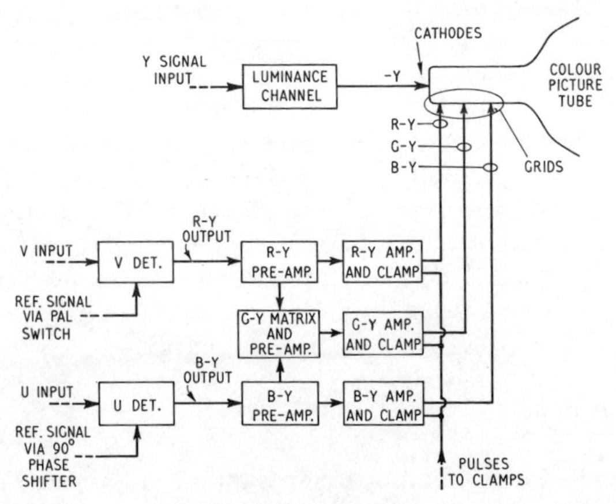

Figure 9.11. Block diagram of video section based on colour-difference drive

Video section

The colouring section concludes with the colour-difference amplifiers feeding colour-difference signals to the grids of the picture tube, and this scheme, called colour-difference drive, is represented by the block diagram in *Figure 9.11*.

The outputs from the V and U detectors are fed respectively to the R−Y and B−Y colour-difference preamplifiers, where the necessary 'de-weighting' of the signals is commonly achieved. So far, of course, we have only obtained the R−Y and B−Y colour-difference signals. The missing G−Y signal is obtained by matrixing the R−Y and B−Y signals, and this often happens in the G−Y preamplifier.

G−Y matrixing

The basic G−Y matrixing function can be understood by reference to the simple network in *Figure 9.12*. Resistors R1 and R2 merely add proportions of the R−Y and B−Y signals to give the correct G−Y signal. To appreciate why the G−Y just seemingly appears from two entirely different signals we need to recapitulate on the composition of the luminance signal. It will be recalled that 100 per cent Y signal is equal to 0.30R+0.59G+0.11B which, of course, is equal

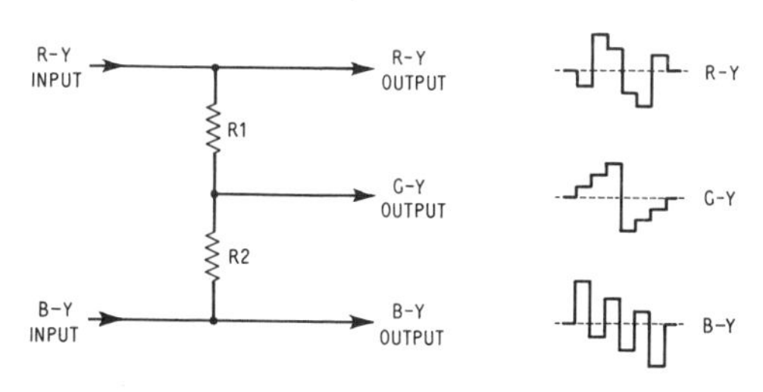

Figure 9.12. Basic G−Y matrix and output waveforms. See the text for details of how the matrixing occurs

to $0.30Y+0.59Y+0.11Y$. Now if we subtract the second expression from the first expression we get $O = 0.30(R-Y) +0.59(G-Y)+0.11(B-Y)$, which gives $-0.59(G-Y) = 0.30(R-Y)+0.11(B-Y)$ and from which is obtained $-(G-Y) = 0.30/0.59(R-Y)+0.11/0.59(B-Y)$. Thus the two resistors in *Figure 9.12* merely add together the R−Y and B−Y signals in the proportions of 30/59 and 11/59 respectively. That is all there is to it really! To help with the understanding of this action waveforms are shown of the three signals during the standard colour bars.

The preamplifier stages are usually transistorised, while the amplifier stages and clamps which connect to the grids of the picture tube may use valves, though recent receivers use transistors here (or in the primary colour drive system—see later).

Grid clamping

The preamplifiers drive the amplifiers so that colour-difference signals of suitable levels appear at the tube grids. Now, an important function here is that of 'clamping'. The waveforms in *Figure 9.12* imply that the datum is through the centre of each (shown by the dotted lines), and this datum has to be clamped to a specific d.c. level to prevent the bias of the guns from drifting and to make the system workable on monochrome signals.

Each colour-difference amplifier, therefore, has associated with it a clamping circuit which is activated by line pulses at the line sync/burst periods. The action is such that each grid always 'sees' a specific d.c. potential during the whole of each line period.

Indeed, on monochrome the video signal at the tube cathodes operates relative to the clamping potential at the grids.

On colour the colour-difference signals at the grids operate above and below the clamped datum and it is this action, as earlier chapters have shown, which produces the colour on the luminance high-definition picture background.

On monochrome, of course, the luminance channel supplies the picture information to the three cathodes simultaneously, the grids then being at a fixed d.c. potential. However, on colour each gun 'sees' the modulation appropriate to the original primary colour signal. This is because the

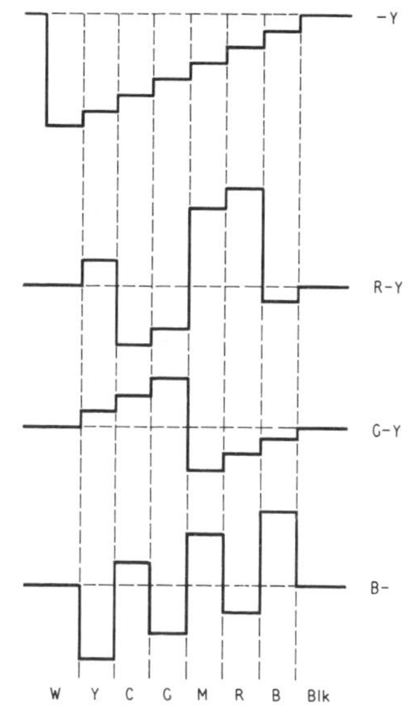

Figure 9.13. The − Y signal is applied to the tube cathodes and the colour-difference drive receiver

guns effectively matrix the −Y and colour-difference signals. For example, the −Y signal (−Y at the cathodes is equivalent to +Y at the grids) is added to each of the colour-difference signals so that we get R from (R−Y)+Y, G from (G−Y)+Y and B from (B−Y)+Y.

The waveforms in *Figure 9.13* are of the −Y signal at the cathodes and the R−Y, G−Y and B−Y signals at the grids due to the standard colour bars, with the levels corresponding to the various colours and black and white indicated. When the Y signal is matrixed with each of the colour difference signals each gun 'sees' its appropriate primary colour signal, as shown in *Figure 9.14* for the standard colour bars. These signals are the equivalent of those delivered by the colour camera at the studio.

Primary colour drive

In some of the more recent receivers the matrixing in this area is handled by a circuit prior to the picture tube, the grid then being returned to a stabilised or controlled potential and the cathodes each receiving the appropriate primary colour signal. This is called primary colour or RGB drive for obvious reasons.

The block diagram in *Figure 9.15* illustrates this method of picture tube drive. The net result, of course, is exactly the same as that obtained from colour-difference drive. The fundamental difference being that in the latter the picture tube provides the matrixing, while in the former a separate circuit (often a part of an integrated circuit) is used.

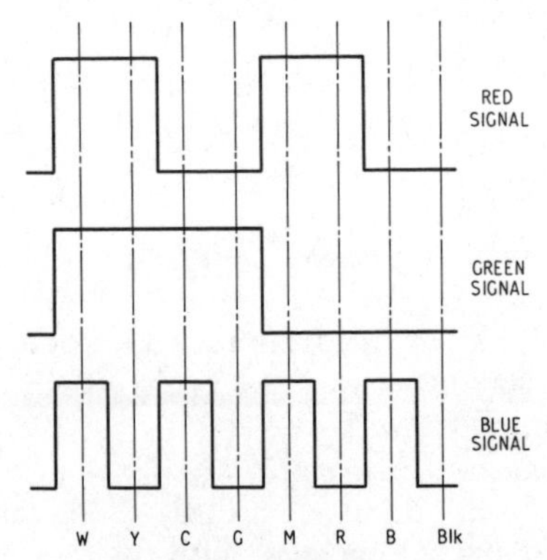

Figure 9.14. The primary colour waveforms resulting from matrixing of the Y and colour-difference signals

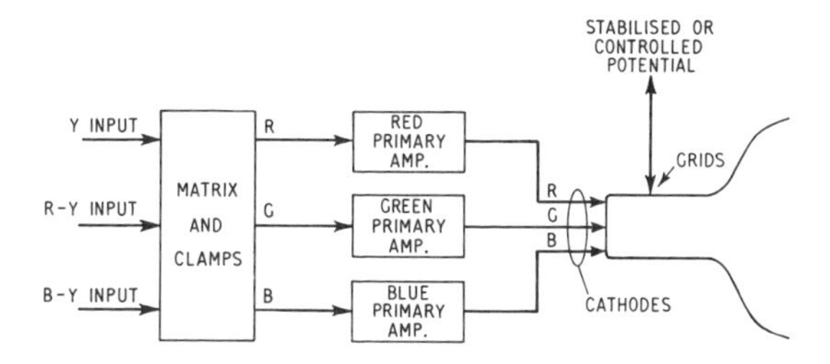

Figure 9.15. Basic primary colour drive system, where the colour-difference matrixing is performed prior to the picture tube. The matrix and clamps section includes the G—Y matrix and may constitute a section of an integrated circuit

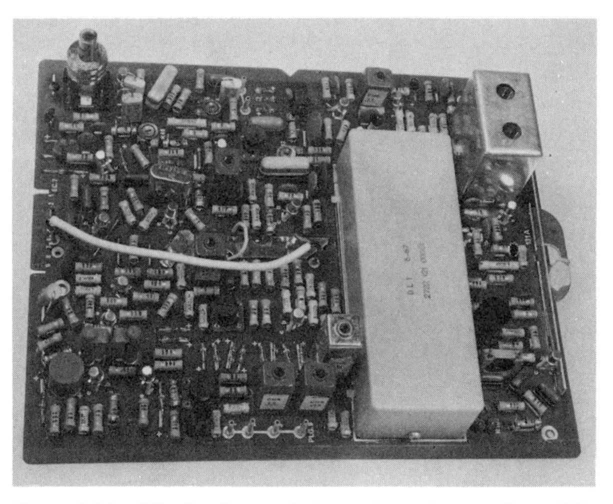

Figure 9.16. The decoder panel of a receiver using transistors. The PAL delay line is the large 'box' on the right-hand side of the panel

10

SECAM BASICS

We have seen that the PAL system differs from the American NTSC system in the following ways.

(1) The subcarrier upon which the R−Y signal is modulated is phase-reversed line by line, so that on alternate lines the V chroma phase is opposite to the phase of the corresponding NTSC chroma signal. Thus the phase on one line is the same as NTSC and on the next line it is reversed, and so on.

(2) The phase of the burst is also switched line by line, so that on the odd lines, corresponding to the phase of the NTSC lines, it is transmitted at 135 degrees, while on even lines, corresponding to the PAL reversal, it is transmitted at 225 degrees. Thus, relative to the −U chroma axis, which is at 180 degrees, it swings in phase over ±45 degrees, line by line. This provides the necessary information on the phase of the V chroma line.

(3) The chroma signal is compounded of V and U components, instead of the NTSC I and Q components, with equal bandwidth transmission of a value similar to that of the NTSC I component, and with specific PAL weighting.

(4) The PAL system also adopts a slightly different subcarrier frequency from the NTSC system for reasons of compatibility.

(5) At the receiver, separation of the PAL V and U components is achieved by delayed and non-delayed lines of chroma signal being added and subtracted in a matrix, the

former arriving by way of a delay line; the result being effective cancellation of hue and level-dependent phase errors and the elimination of the NTSC hue control.

The French SECAM (short for Séquential couleur à mémoire) system (adopted by France, Algeria, German Democratic Republic, Hungary, Tunisia and the USSR) differs from both NTSC and PAL mainly in the following ways:

(i) The colour information is transmitted sequentially—e.g., one after the other in time.
(ii) The chroma components are derived from the R—Y and B—Y signals frequency modulating the subcarrier.

One advantage lies in the fact that the attributes of f.m. (e.g., immunity from amplitude interference, good signal-to-noise performance, etc.) are bestowed upon the colouring system. Also of prime importance is the fact that since the colouring information is transmitted sequentially—only one colour-difference component at any one time—the SECAM receiver requires no auto-phase control loop on the reference signal generator or, indeed, synchronous chroma detectors. It is thus less complicated than an NTSC or PAL receiver, but it needs a delay line.

It is a compatible system in which the luminance signal is transmitted as amplitude modulation of the vision carrier.

There is no reliance upon the phase of the chroma signal for hue reproduction, as there is with NTSC and PAL, and thus like the PAL system it offers immunity against phase distortion, but for a different reason. Also, unlike NTSC and PAL, it can suffer no disturbance due to crosstalk between the two chroma channels.

Total SECAM bandwidth is equal to that occupied by a monochrome transmission using the same line standard.

SECAM, though, has a reduced vertical resolution compared with NTSC, but this is only of minor subjective consequence. Possibly one of the most adverse of subjective effects is the likelihood of flicker along the horizontal edges of a display, especially between coloured areas of high saturation, at half field frequency.

There are other comparative and inter-related technical factors between the three main colour systems, but it falls outside the brief of this book to examine them (see, for example, the report of the EBU Ad-hoc Group on Colour Television, 1965). It was after highly tutored deliberation that the PAL system was chosen for the U.K. on 625 lines.

The r.f. and i.f. stages are the same for the three types of receiver; also once the colour-difference signals have been recovered the receivers are again identical. It is only a portion of the receiver which differs in the three cases—essentially that concerned with the decoding of the colour information.

Chroma demodulation

It will be recalled that the sound of the 625-line television system is frequency modulated and that a special type of demodulator or detector is required to extract the audio signal from the carrier wave. The same sort of circuit is used in the SECAM decoder. The subcarrier signal is a sine wave, the frequency of which is changing by a number of cycles determined by the degree of modulation. This is applied to the detector input, and the output is then a voltage whose value is determined by the amount of frequency change and hence the degree of modulation. Thus a carrier frequency variation or modulation is converted into a changing voltage and is therefore demodulated. It will be understood that while the amount of frequency change (deviation or modulation depth) governs the amplitude of the demodulated signal, the *rate of change* governs the frequency of the demodulated signal.

The other special components needed in the SECAM receiver are a delay line and an electronic switch. The delay line—as used in the PAL system—holds up the first part of the colour information so that it can be presented simultaneously at the pair of f.m. detectors. The electronic switch assists with this action. It is operated by pulses from the line timebase circuit, this being possible because the sequential periods are about equal to the time taken to scan one line on the picture tube screen.

SECAM operation

Figure 10.1 gives a simplified block diagram of the SECAM receiver. Since the subcarrier is not suppressed at the transmitter, as it is with NTSC and PAL, there is no need for the reference generator circuits and control to re-form it at the receiver, and the point where the chroma signal is communicated to the detector is the point at which we take up the story of the SECAM receiver.

The following description will be referred to the I and Q chroma components of NTSC. SECAM has equivalents, of course, which are based on the R−Y and B−Y colour-difference signals, sometimes called D_R and D_B, corresponding respectively to the R−Y and B−Y axes, where

$$D_R = -1/0.7(R-Y) \text{ and}$$
$$D_B = -1/0.89(B-Y)$$

When the I and Q information has been released and it is about to rejoin the Y (luminance) signal, then the story is again the same for both NTSC and SECAM. We should bear in mind that NTSC alters the R−Y and B−Y signals to I and Q signals respectively due to the nature of the suppressed subcarrier modulation.

Now, how the SECAM receiver works between the two points mentioned can be explained as follows. Suppose at the instant we are looking at the block diagram shown in *Figure 10.2*, the SECAM system is sending the I information, then the electronic switch will be positioned so that this information (I_1) is sent direct to the I detector. The Q detector is receiving the previously sent information which,

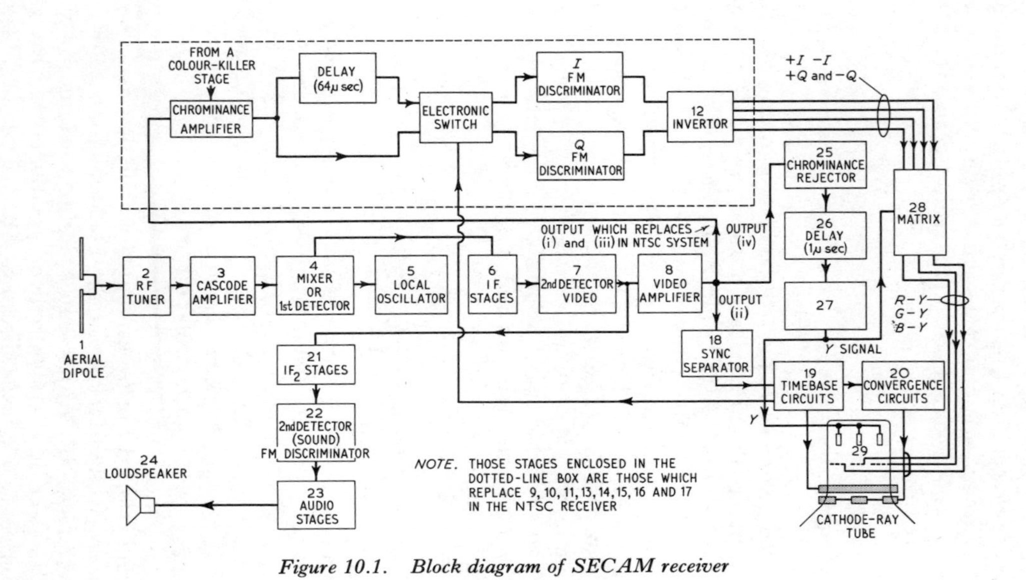

Figure 10.1. ***Block diagram of SECAM receiver***

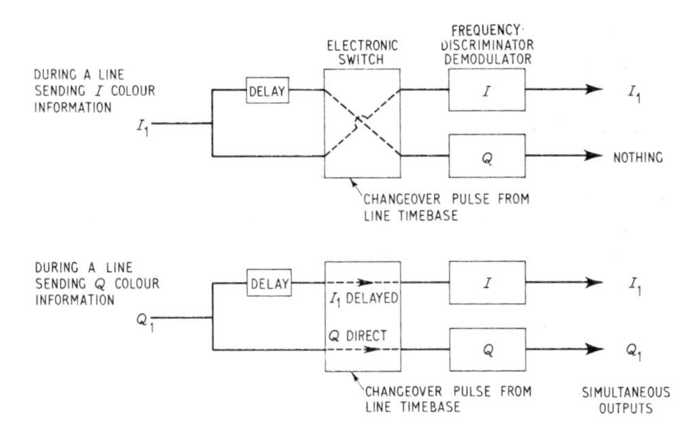

Figure 10.2. Block diagram showing sequence of operations in a part of the SECAM decoder

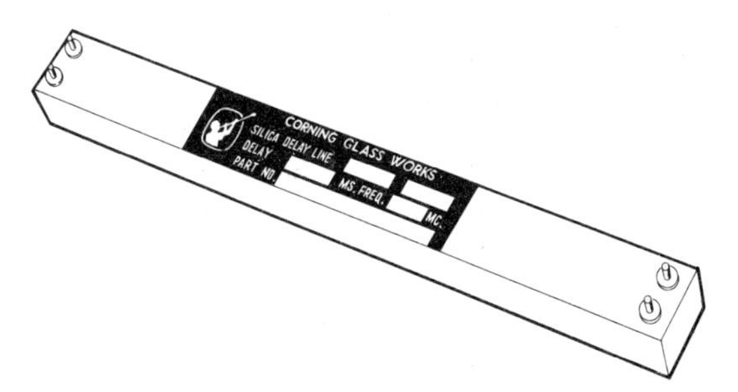

Figure 10.3 SECAM Delay Line (drawing from a photograph by courtesy of Corning Glass International S.A.)

since this is assumed to be the instant of switching on, is nothing at all! The switch now changes over and the Q signal (Q_1) arrives direct and the previous I_1 signal, which has been delayed (since there is a common input to the circuits), now emerges (after a 64 μs delay provided by the delay line) to join the Q_1 signal. The changeover is repeated, giving next I_2 Q_1 (again), Q_2 I_2 (again), I_3 Q_2 (again), and so forth. It will be seen that each colour information signal is used twice, once direct and then delayed, and in consequence some of the vertical definition of the picture is lost, but this is barely discernible under normal viewing conditions, as already noted.

The electronic switch

The switch is operated by pulses from the line timebase, since the period used for sending each 'packet' of colour information is a line period—that is about 64 μs.

In practice it is possible for the switch to operate so that each colour sequence is being fed to the wrong detector. To avoid this a special sawtooth waveform is transmitted at the conclusion of each field sync pulse which operates a polarity-conscious diode pulse-yielding circuit. This synchronises the switch action to the colour information. It will be recalled that a similar identifying function is built into the PAL system in the form of the swinging bursts.

11

COLOUR RECEIVER CONTROLS

An impression of the main controls on a colour receiver is given in *Figure 11.1*. The actual disposition of the controls differs between receivers and the hue control shown is characteristic of NTSC receivers and of non-PAL receivers

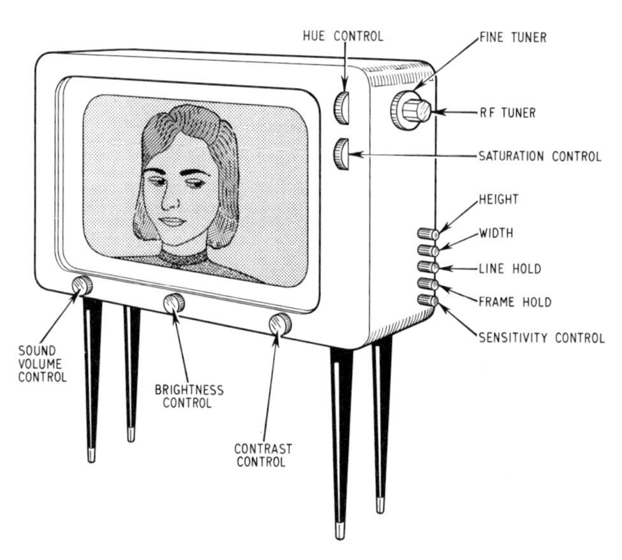

Figure 11.1. Outline of some of the controls of a colour receiver (other internal and preset controls are detailed in the text)

working from a PAL signal. In the latter respect, so as to avoid infringing the PAL patents, a few imported receivers are designed such that the PAL signal is translated to a form of non-PAL which is then accepted by the receiver and reproduced rather after the style of an NTSC signal without the PAL attribute of colour hue 'lock'. Hence the need for a hue control.

Moreover, some real PAL receivers are also equipped with a 'hue control' (sometimes called 'tint control'), but this differs from the NTSC control of this name since its purpose is not to correct for phase distortion on the signal but to allow the viewer to adjust the display slightly either towards magenta or cyan to match his subjective impression of colours. The hue control is commonly adjusted for the realistic reproduction of flesh colours.

Main controls

Colour receivers are equipped with two sets of controls, the main controls shown in *Figure 11.1*, which are generally available to the user, and preset controls which are located mostly inside the receiver and require expert adjustment for optimum results. First, let us deal with the main controls.

Hue control (NTSC)

Should the system or signal propagation conditions cause a change in phase of the chroma signal then the colours in an NTSC display will change from their correct hues. The NTSC hue control provides a means of manually correcting such phase error, thereby enabling the colours to be restored towards their natural value, either by swinging the quadrature colour axes as a whole a number of degrees either side of normal, relative to the regenerated subcarrier (reference signal), or by swinging in a like manner the phase of the reference signal relative to the colour axes. In either case a change in colour reproduction results, and it is left to the viewer to decide for himself which control setting yields the most natural display.

Tint control

This control, not shown in *Figure 11.1*, is found on some PAL receivers. Its function is not to alter the phase of the chroma or reference signal, but rather to regulate differentially the colour-difference drives to the picture tube (or the biasing) so as to secure a change in tint (or white level).

It will be fully understood by now, of course, that PAL receivers do not require a hue control because the displayed colours are effectively 'locked' to the transmitted hues by the PAL function. The tint control on PAL receivers, therefore, is more a refinement than a necessity.

Saturation (or colour) control

This control, used on all colour receivers, merely regulates the amplitude of the chroma signal fed to the chroma detectors and thus adjusts the 'intensity' or saturation of the colours reproduced on the high-definition luminance background. When this control is fully retarded many receivers give a normal display in monochrome only; but if the control is fully advanced then the colours will be far too bright and unnatural. Viewers first experiencing colour television are inclined to set this control for unnaturally bright colours!

Main controls common to both monochrome and colour receivers

Many of the main controls of colour receivers have counterparts on monochrome receivers and for optimum adjustment of these and, indeed, the colour controls, a test card display is desirable. Test Card F, shown in the *Frontispiece*, is designed for assisting with the adjustment of both monochrome and colour receivers, though at the time of writing a different test card is undergoing trials in the London transmitter areas. Let us now look at the common controls and see, where applicable, how Test Card F can help with their adjustment.

Volume control and tone controls

These controls merely affect the sound reproduction and have nothing to do with the picture. The volume control merely regulates the amplitude of the audio signal and hence the loudness of the sound reproduction.

Not all receivers incorporate a tone control or tone controls, but when there is a single tone control this generally tends to diminish the treble output as it is advanced, giving a 'mellow' sound which, although incorrect, is favoured by some people. Receivers with more advanced audio sections may feature two controls for tone which, in good designs, provide boost and cut of both bass and treble frequencies. In general, however, the audio sections of television receivers are incapable of 'hi-fi' reproduction owing to the relatively high amplifier distortion and the diminutive loudspeaker units and their nature of loading in the cabinets. However, TV sound tuners and adaptors are appearing which make it possible to extract the audio signal prior to the audio section for communication to a hi-fi system. Some receivers are also designed with special attention to the sound reproducing department.

Brightness control

This regulates the amount of bias applied to the three guns of the picture tube and hence the amount of beam current (e.g., the quantity of electrons emitted from the guns and sent in terms of electron beams to the screen).

This adjustment is best made in conjunction with the contrast control (see below) and with the colour control fully retarded. The degree of brightness required will depend on the ambient illumination and the general viewing conditions; but if it is attempted to obtain excessive brightness, to combat high room illumination, for example, the display may tend to go out of focus and the peak white parts may be clipped due to the action of the beam current limiting circuit, which most recent receivers incorporate.

Contrast control

This control regulates the amount of luminance signal presented to the picture tube, either directly to the cathodes, or to the primary colour matrix. It should be adjusted in conjunction with the brightness control to give the correct relative display of the six contrast rectangles to the left of the centre circle of Test Card F. These have a contrast range of about 30:1 from almost white to black (top to bottom). When correctly set up the small lighter spots within the top and bottom sections should just be visible.

Tuning control

The nature of tuning or channel selection is determined by the design of the receiver. Some models have a tuning control knob generally calibrated in u.h.f. channel numbers, while others, particularly of more recent design, feature push-button or 'touch' button channel selection.

The tuning or channel selection of a colour receiver is more critical than that of a monochrome receiver, though the principle remains the same. The idea is to tune first for optimum luminance definition, consistent with buzz-free sound reproduction, and then, with the colour control advanced, to optimise the tuning if necessary for the least pattern effect which can result from interference between the chroma and sound signals when the control is turned very slightly beyond the optimum definition point. It is sometimes best to tune until the chroma/sound pattern occurs, and then to 'detune' *slightly* away from this point until the pattern disappears and the sound is optimised.

The nature of the tuning used with push-button channel selection is a function of the design of the u.h.f. tuner; but each button is tuned when depressed either by an associated or common control for the required channel, the tuning technique being the same as just described.

It is noteworthy that if the tuning is adjusted incorrectly a monochrome display may result, the colours appearing only when the tuning is optimised.

Height and width controls

On some receivers these controls may be at the rear or inside the cabinet as presets. They regulate the vertical and horizontal power in the corresponding scanning coils on the tube neck. In general, the height control should be adjusted so that the visible picture height extends from the top line of the top colour bars of Test Card F to the point of the bottom arrowhead, after which the width control should be adjusted until the side castellations of the Test Card appear in the display area of the screen. It is noteworthy that the transmitted aspect ratio (width-to-height ratio) is 4 : 3 while some picture tubes have a display aspect ratio of about 5 : 4. However, when the height and width are adjusted as described the lack of exact coincidence is masked.

It may be necessary to shift the picture vertically or horizontally on the screen and colour receivers incorporate preset controls for this purpose (see later).

Vertical and horizontal hold controls

These controls may also be preset, and some recent timebase designs eliminate the need for manual controls of this type. However, when fitted their purpose is merely to adjust the field and line timebase repetition frequencies so that the display synchronises exactly with the transmission. When the vertical hold control is maladjusted the picture will 'roll' either up or down as though it were rotating on a drum inside the receiver. After adjusting for vertical lock it may still be necessary carefully to adjust the control within the locking range to optimise the interlacing of the picture— until the lines of the two fields which go to compose a complete picture properly interlace.

Maladjustment of the horizontal hold control results in horizontal picture tearing and side-slip, so that it may be impossible to see anything except a jumble of incoherent lines.

Test Card F reveals faulty line synchronism by horizontal displacement of the castellations of the side borders.

Preset controls

It is generally undesirable for the non-skilled viewer to attempt any adjustments to internal controls. Not only could this put the receiver grossly out of adjustment, eventually demanding skilled attention, but it could also prove dangerous if not lethal owing to the 240 V mains potential and the 25 kV of e.h.t. present inside the receiver.

However, it is sometimes possible to adjust certain preset controls from the outside of the receiver, without the need to remove the rear cover, and these might include the controls for vertical and horizontal shift, dynamic convergence, first anode potentials of the picture tube for partial grey-scale tracking, etc. Manufacturers are tending to arrange for the shift and convergence controls to be grouped on a panel (sometimes removable) which is accessible from the front of the receiver. The alternative is to use a fairly large mirror so that the effect of the adjustments can be seen from the rear of the cabinet.

Shift controls

These merely move the complete picture horizontally and vertically on the screen for centring. The service technician may deliberately shift the picture downwards to observe the colour bars at the top of Test Card F, restoring the adjustment afterwards, of course.

It may be necessary to make adjustments to the height and width in conjunction with the shift controls, and if the picture geometry is incorrect the vertical and horizontal linearity controls might also need adjusting at the same time.

Linearity controls

There is generally such a control for horizontal linearity (sometimes a core movement in a coil) and always at least one for vertical linearity. These are best adjusted while observing the inner circle and the background squares of Test Card F. The idea of the adjustments is to secure the most symmetrical circle and properly dimensioned squares.

When there are two vertical linearity controls, one commonly adjusts the linearity of the whole picture in the vertical sense, while the other might affect essentially the top or bottom of the picture, usually the former.

Other controls for initial setting up

For initial setting up it may be necessary to adjust the orientation of the picture by rotating in the appropriate direction the scanning coil assembly on the tube neck, after slackening the clamp screw.

Colour receivers also usually incorporate a focus control which regulates the potential on the focus electrodes of the picture tube. This should be adjusted for the best focus of the scanning lines.

Some models are equipped with an adjustment for correcting pin-cushion distortion of the display. When fitted, this is adjusted for the least bowing of the edges of the picture.

There are also hosts of additional preset controls for adjusting such things as the e.h.t. potential, the beam current limiter operating threshold, tuner a.g.c. delay, PAL matrix, decoder functions and so on, but these should never be adjusted without full knowledge of the receiver or without reference to the service manual.

One adjustment common to all receivers, which should be checked upon initial installation, is that for the mains voltage, but even some of these adjustments are rather complicated and require reference to the service or installation manual. The nominal U.K. mains is 240 volts, and most receivers are set to this voltage when dispatched.

Degaussing

All recent colour receivers incorporate automatic degaussing which operates each time the receiver is switched on. While this normally takes care of any magnetic effects induced into the shadowmask and associated metalwork, an external degaussing coil may be required to eliminate more persistent magnetic inductions. It is desirable to apply external degaussing prior to purity and convergence adjustments.

Purity

The principles of purity are expounded in Chapter 7, so there is no need to repeat them here. Adjustment is made by ring magnets on the tube neck until a raster yielded with only one gun operating is pure in colour. The red gun, and hence the red raster, is the commonly adopted colour, since impure red is readily discerned; however, it may be necessary to check on the other colours and make fine compensating adjustments on each of these, returning to red again for a final check.

Convergence

Chapter 7 also examines the need for convergence, revealing how a colour picture element results from the registration of the components of the element in the three primary colours —red, green and blue. Lack of registration means that the component colours fail to overlap; they thus show separately with displacement between them.

Centre-of-screen convergence (static convergence) is achieved by fixed magnetic fields operating separately on each beam, while the convergence away from the deflection centre (dynamic convergence) is handled by changing magnetic fields produced by electromagnets energised from specially tailored signals from the line and field timebases. Again, each beam has its own set of coils, and controls are present so that the energising currents can be adjusted for the best convergence over the whole area of the screen.

Some models have the dynamic convergence adjustment procedure printed near to or on the panel carrying the convergence controls; but it is generally necessary to refer to the service manual for full information about the adjustments, as the method of adjustment can differ between receivers.

To assist with the checking of the convergence, certain grid lines of Test Card F are outlined in black, and it is here where colour misregistration will show when the convergence is in error. However, should apparent misregistration occur on the lines without the black edging, then the trouble could be caused, not by misconvergence, but by low-frequency ringing or distortion in the chroma circuits.

The blackboard and white cross of the picture in the circle provide a check on static convergence.

A crosshatch and dot generator is required to secure the best convergence possible by the receiver design, and this often provides a blank raster for purity adjustments. Even after the convergence has been carried out correctly, it is not abnormal for mild misconvergence to be observed close to the screen, particularly at the extremes of the display, but the effect should not be apparent at normal viewing distance.

Grey-scale tracking

This set of adjustments is to ensure that the currents of the three beams change in unison with changes in brightness of the picture, and that the correct ratio of change occurs under dynamic conditions. They are needed to 'equalise' the characteristics of the three guns of the picture tube.

At the low brightness end of the range the equalising is achieved by adjusting the cut-off characteristics of the three guns by the first anode potentials. A control for each gun is thus provided, and the adjustment is best made in a darkened room so that the cut-off threshold can be observed on the display. At the high brightness end of the scale, adjustment is provided in some receivers by the video 'drive' presets.

When the characteristics of the three guns coincide at the two ends of the scale, reasonable tracking over the range of brightness control can be expected. Incorrect tracking can cause the greys in a monochrome display to show colour tint, the degree of contamination tending to change as the brightness control is adjusted. The tracking also corrects for the differences between the efficiencies of the red, green and blue phosphors of the screen.

White level

The nature of the white displayed by a 'white' raster will depend on the energies of the three beams and hence on the red, green and blue lights produced. For example, too much blue light will swing the raster towards blue—and similarly with the other two colours. The white value chosen for colour television is called Illuminant D, having a bias slightly towards green (the earlier-used Illuminant C was slightly towards magenta).

Some receivers have controls marked red, green and blue adjustable from the rear to balance up the white level, and devices are available for providing Illuminant D light as a comparative.

Other features of Test Card F

We have already seen some of the checks that Test Card F provides. Others are given below.

Colour bars

After the conclusion of the field blanking period in each field, the top of colour Test Card F displays a few lines of electronically modulated standard colour bars of white, yellow, cyan, green, magenta, red, blue and black of 100 per cent amplitude and 95 per cent saturation.

These are useful for revealing any shortcomings in the performance of the colouring circuits, and they can also be used for assessing the performance of the decoder by switching off the individual guns of the tube.

Top border castellations

These are in high luminance cyan and they should normally be displayed in constant saturation. Variations in saturation could signify that the reference generator is failing to recover in phase sufficiently quickly over each line.

Bottom border castellations

These are in green, and if they show saturation variations, then a shortcoming could exist in the reference generator control referred to the ends of the fields, relative to the starts.

Left border castellations

These are in red and blue, and if the burst gate lets through picture information or is working wrongly, then violent changes in saturation or Venetian blind interference is likely to occur across the display, in line with a castellation or castellations.

Right border castellations

These, in yellow and white, give an idea of the performance of the sync circuits with or without subcarrier respectively. The effect when the circuits are in error is variation in picture content stemming from the appropriate castellation.

Black and white vertical lines of the grid

Apart from showing up misconvergence, as we have already seen, these lines also reveal multipath interference and signal reflections in the form of 'ghost' lines.

Black and white corner stripes

These show how well the receiver focuses at the corners of the screen.

Sets of vertical lines at right of circle

From top to bottom these correspond to video information of frequencies 1.5, 2.5, 3.5, 4.0, 4.5 and 5.25 MHz. The gratings thus signify how well the receiver defines information of increasing detail. Most modern receivers resolve up to the 4.5-MHz gratings at least.

Top 'letter box' pattern

This is the black rectangle within a larger white rectangle. If the display shows streaking at the right edges and corresponding border castellations, then the low-frequency video performance of the receiver could be in error.

Contrast gradations at left of circle

This column of six rectangles corresponds to an overall contrast ratio of about 30 : 1, and the brightness difference between adjacent rectangles should be fairly constant on a well-designed and correctly adjusted receiver. The small lighter spots in the top and bottom rectangles show whether white or black 'crushing' is occurring, by the spots merging into the surrounding areas.

Colour picture in circle

This gives an overall appraisal of the colour display. The child model instead of an adult is deliberately used to avoid dating of the card with fashion changes, etc. The flesh tones and other colours are useful for discerning the effect of a tint control, when fitted.

INDEX